7-88

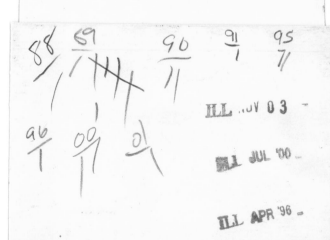

7-88

HANDBOOK OF
ELECTRICAL NOISE
MEASUREMENT AND TECHNOLOGY

2ND EDITION

CHARLES A. VERGERS

 TAB Professional and Reference Books

Division of TAB BOOKS Inc.
P.O. Box 40, Blue Ridge Summit, PA 17214

621.38/53

FIRST EDITION
SECOND PRINTING

Printed in the United States of America

7-88   BT   4000

Library of Congress Cataloging in Publication Data

Vergers, Charles A.
    Handbook of electrical noise.

    Includes index.
    1. Electronic noise. 2. Electronic circuits—Noise.
I. Title.
TK7867.5.V47 1987   621.381'53   86-23132
ISBN 0-8306-2802-9

Questions regarding the content of this book should be addressed to:

    Reader Inquiry Branch
    Editorial Department
    TAB BOOKS Inc.
    P.O. Box 40
    Blue Ridge Summit, PA 17214

Contents

Experiment 9 Bandpass Filter—Experiment 10 Notch Filter—
Experiment 11 Compressor and Expander Circuits—Experiment
12 Design of Transistor Voltage Amplifier Using Optimum Collector
Current

Introduction

This is an introductory book on the subject of electrical noise. I teach electronics engineering technology courses at Capitol Institute of Technology, a four year college of engineering technology located in Laurel, Maryland. One of the courses that I have instructed many times at the College is Noise. After being unsatisfied with the types of books that were available for use in the course, I started with the purpose in mind to write a book which would be suitable for an undergraduate course in electronic engineering technology, as well as being a general information book on the subject of electrical noise.

When I teach a course I believe in making things as clear as possible for the learning process. Too many times I have seen books with derivations as well as problems presented with many missing steps. This makes it very hard for someone who is attempting to learn a subject for the first time. In this book you will see complete derivations, and copious example problems worked out completely. Also, I have included a group of problems at the end of each chapter. In the new edition I have included solutions to the problems at the end of each chapter. Also, I have written a comprehensive exam with answers. After finishing the book you can take the exam to test your knowledge of the subject.

This book begins with a chapter that introduces and classifies the various types of electrical noise. It also gives a number of statistical terms necessary for the understanding of an electrical noise process. Following this is a chapter that deals with the tools necessary to solve noise problems. New material in this edition includes the Fourier Transform and correlation problems.

Chapters 3 and 4 bring up the subject of thermal and electronic device noise, respectively. Next comes the subject of noise in advanced networks. A more in-depth treatment of noise equivalent bandwidth is given in the exercises than in the first edition. Next comes noise figure and noise equivalent temperature, just prior to noise measuring instruments and equipment. A chapter is devoted to noise measurements, and the chapter on noise in communication systems has been rewritten to include pulse modulation and pulse-code modulation. A chapter is devoted to low noise circuit design.

More experiments have been added and the book has many drawings to help the various explanations. A minimum amount of calculus is needed to understand this book. Someone without a calculus background could elect to study those subjects not requiring the calculus. The majority of the calculus is used in noise equivalent bandwidth subjects.

This book should be suitable for an audience made of either associate or bachelor degree program students in electrical engineering technology. Also, the book is a practical reference for those individuals interested in electronics who require information on electrical noise for either their occupation or general interest.

I would like to express my appreciation to the following individuals for their help in the preparation of this manuscript. Mrs. Linda Lauer formerly of our college for typing portions of the manuscript. Miss Katherine Feldmann and Mr. Mark Yembrick, two of my former students, for proofreading parts of the manuscript. Mrs. Ollie Harmon of TAB BOOKS Inc., for typing and photocopying portions of the manuscript. Also, many thanks to TAB BOOKS Inc. for allowing me to prepare the second edition of this book.

Chapter 1

Noise Types, Terminology, and Statistical Terms

T HERE IS ONE SUBJECT WHICH EVERYONE WHO WORKS IN electronics becomes familiar with, sooner or later. That subject is electrical noise.

Noise is any signal which interferes with the transmission of a signal through a network or tends to mask the desired signal at the output terminals of the network.

From our definition of electrical noise, it can be seen that noise can be either periodic or non periodic in nature. However, normally electrical noise is of a random nature which is indicative of a complex process of both periodic and non periodic waveforms.

Noise can be examined outside the electrical system as well as inside the system. An example of noise generated outside of a system would be electrical disturbances due to thunderstorms while inside noise would be that generated by resistors and active devices in the electronic circuitry.

NOISE CATEGORIES

Noise can be broken down into many different categories. I will use three main categories which are erratic, man-made and circuit noise. Erratic disturbances can be divided into two areas known as atmospheric and space noise. Atmospheric noise or static is caused by lightning discharges and other natural electrical disturbances resulting in electrical discharges. Space noise is caused by our sun as well as distant stars.

Man-made noise is the result of machines or ignition systems which give off an electrical spark.

Circuit noise is brought on by resistors, vacuum tubes and transistors. These components generate what we could call a spontaneous fluctuation. It is generated due to thermal effects in each of these components and by an additional process called the shot effect in vacuum tubes and transistors. Figure 1-1 illustrates a noise hierarchy diagram. We must remember that the subject being addressed is electrical noise and not acoustical noise.

Now let us look at each of these types of noise a little more carefully. First of all, let us examine erratic noise. To do this we must know something about static electricity.

Static Electricity

Static electricity deals with the subject of electric charges at rest. The unit of electrical charge is the coulomb. One coulomb is a charge representing the charge due to 6.28×10^{18} electrons. In other words if 6,280,000,000,000,000,000 electrons were present at some point, the equivalent charge would be said to be one coulomb. An absence of these electrons could be considered a positive charge. This is because the electron is negative in electrical charge. Thus we can state that excess electrons constitute a negative charge. An absence of electrons constitute a positive charge. From basic physics we learn that like charges repel and unlike charges attract.

Whenever we examine the forces acting on charges we will find out that:

1. The force depends on the product of the two charges and is proportional to this product.
2. The force is inversely proportional to the square of the distance between the charges.

These facts can be summed up in formula form by Coulomb's Law. It is as follows on page 14.

Equation 1-1 $$F = \frac{(K)\,(Q_1)\,(Q_2)}{D^2}$$

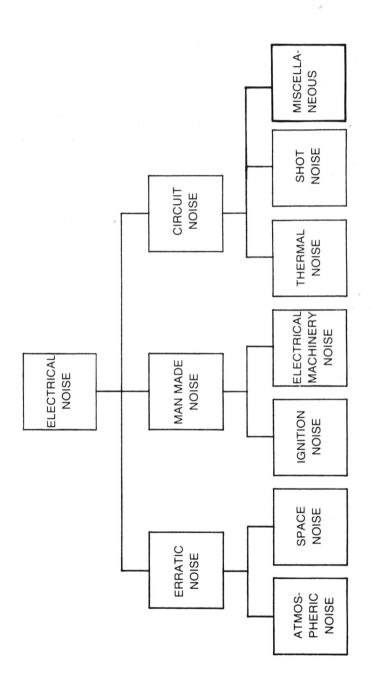

Fig. 1-1. Noise hierarchy diagram.

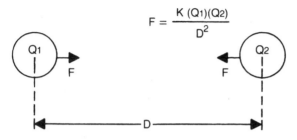

Fig. 1-2. Illustration of the use of Coulomb's law.

The symbol F is the force acting between the charges in newtons and Q_1 and Q_2 are the charges in question. D is the distance between the charges in meters. K is a constant dependent on the medium in which the charges are present. That is approximately 9×10^9 meter—volts/coulomb in free space. In Fig. 1-2 we see an illustration of this formula's use.

There is a more valuable formula for which we are looking. This formula can be used to calculate the potential difference between some point in the vicinity of a charge with respect to ground. Refer to Fig. 1-3. You can see that the voltage in respect to ground depends on the distance from the charge to the point in question, the size of the charge, and a constant K.

The formula is:

Equation 1-2

$$V = \frac{(K)\,(Q)}{D}$$

As an example of the use of this formula, let us say that a charge of 6 mC is located 9 cm from a point in space. Compute the voltage at the point with respect to ground.

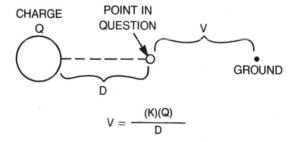

Fig. 1-3. Potential difference at a point in space.

Solution

$$V = \frac{(K)\,(Q)}{D} = \frac{(9 \times 10^9)\ \text{meter-volts}\ (6 \times 10^{-3})\ \text{coulomb}}{\text{coulomb}\ 9 \times 10^{-2}\ \text{meters}}$$

$$= 6 \times 10^8 \text{ volts.}$$

The above information can be used in the discussion of static discharges due to thunderstorms. The most dramatic of all static discharges is that of lightning. During a thunderstorm high winds tend to break up water droplets into various sizes. One theory says that the smaller droplets will become negatively charged and the larger droplets become positively charged. Areas of charge will accumulate in the clouds and when the potential difference is high enough, a discharge will occur. This discharge, which we call lightning, may also occur between cloud and earth. Regardless, this lightning will cause much interference with radio communications. To see why, let us attempt to use the formula given by Equation 1-2 to understand the size of a lightning voltage. Imagine that a group of negative charges are carried high into a cloud during the storm. The size of the voltage that will produce the lightning is dependent on the charge accumulated as well as the distance from the charge to the point of which we are in question. Since the point at which the discharge occurs is very near to the charge, the distance D is quite small and the result is that the voltage is extremely large. In some cases, it is several hundred million volts.

The subject of erratic noise is very important to us because of the interference it causes with radio and television broadcasts. Let us examine the subject of lightning a little more closely.

The earth has been found to be at minus 300,000 volts in respect to the ionosphere. This is due to cosmic rays which are always charging the various atmospheric levels. The result is that electrons are constantly being pulled out of the ground to zoom up into the sky. Thunderclouds get caught in this electrical path. As mentioned before, the water droplets in the clouds become charged or polarized. This is also true of ice crystals. The cloud base usually becomes negative since it has absorbed electrons, and earth under the cloud becomes positive in respect to the cloud bottom. Now, since the storm clouds move, the positive charge on the ground level moves. Also, as time goes by and the storm matures, the charge on the ground increases as the charge in the clouds increases.

This ground charge moves slowly into tall structures, like trees and church steeples. These charges are trying to get to the cloud because of the force we mentioned when we discussed Coulomb's law. Since the ground charge is positive, the cloud charge is negative and unlike charges attract, then we see why the ground charge is trying to get to the cloud charge.

The ground charge cannot easily get to the cloud charge because of air between the cloud and ground is a very poor conductor of electricity. An extremely large voltage must exist between the earth and cloud before a discharge can take place. This voltage may be as much as 100 million volts or even higher.

To understand the actual lightning stroke we must trace the series of events which give us a luminous lightning stroke. First of all, a series of premature strokes called step leaders travel from the cloud downward toward the ground. These leaders are not luminous and last for as much as two microseconds. The leaders may never reach the ground until they establish a well-ionized conductive path through the air. The ground charge moves upward trying to get to the step leaders. When contact is made, the ground charge will move quickly into the cloud and the surrounding air molecules are heated to blazing incandescence. The result is that we see lightning. The visible part of the lightning is referred to as the return stroke since it travels from the ground toward the cloud. The return stroke is usually complex and may be made of many strokes. Usually four surges are common and they last a total of 0.2 seconds. Lightning bolts have been known to range in length from about two-tenths of a mile to 100 miles. The most typical bolt is about one mile long. Energy content may be equivalent to 400 horsepower. The amount of energy content from all of the lightning strokes reaching the earth in one day is equivalent to about 17 million lightning strokes, each of 400 horsepower energy content. Regarding the speed of the return stroke, a speed of over 60,000 miles per second is typical (Fig. 1-4).

Due to the violent change in the air pressure as the air explodes in the lightning channel, we have acoustical noise which we call thunder. However, the gigantic spark we call lightning causes another type of noise. The noise is electrical.

The electrical nature of the lightning pulse can be examined by drawing a model of an induced voltage in a wire due to lightning. The lightning induced voltage reaches a maximum in about 1.2 microseconds, and then decays from this maximum to 50 percent of the maximum in about 48.8 microseconds. Figure 1-5 illus-

Fig. 1-4. Completion of conductive channel between ground and cloud.

trates this wave. About 150 microseconds later the pulse has decayed to near zero volts.

If we make a frequency analysis of the induced voltage we would find that the frequency spectrum would fall off slowly with frequency. A very broad estimate for the frequency spectrum is as shown in Fig. 1-6.

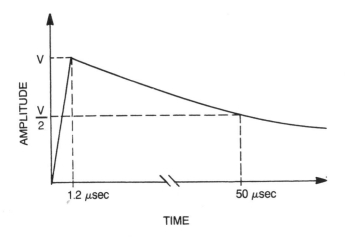

Fig. 1-5. Model of voltage induced in a wire due to lightning.

It has been estimated that there are about 2000 thunderstorms taking place on earth at every moment. You will receive disturbances from these thunderstorms, on radio or television. The reason is that the frequency makeup of the lightning pulses overlap the frequency allocation of the radio-television transmission spectrum. It is obvious that these disturbances from local thunderstorms will be more severe but less numerous than that of storms taking place at other locations on the earth. Also, static will be more severe at night from distant thunderstorms. Above 30 megahertz (MHz) transmissions are limited to the line of sight range and the

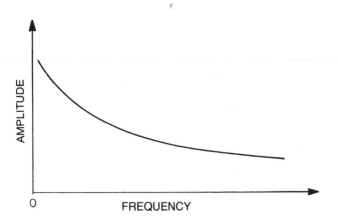

Fig. 1-6. Frequency spectrum appearance for induced voltage due to lightning.

electrical noise due to lightning will be of little effect. Thus, a television will be less affected than a radio from the lightning's electrical makeup.

Space Noise

This form of erratic disturbance is caused by our sun as well as distant stars. There is a constant radiation from our sun as well as a cycle of electrical disturbance of corona and sunspots. This cycle repeats itself every 11 years with a supercycle which seems to be increasing in strength every 100 years.

Distant stars generate noise. This noise is sometimes called blackbody or thermal since it is due to the high temperature of these stars. This noise is uniform over a considerable part of the frequency spectrum.

It may be mentioned in passing that we are also receiving noise from other galaxies. Space noise can be measured from about 8 MHz up to 1.5 gigahertz (GHz).

The branch of astronomy known as radio astronomy deals with the study of electromagnetic radiation from the various celestial bodies. The signals received from these bodies are no longer considered noise by many astronomers since they indicate information about celestial bodies many light-years from earth.

Man-Made Noise

Whenever an electrical machine or ignition system produces a spark or radiates some undesirable electrical wave we have man-made noise.

For example, an ignition coil and distributor system may cause serious electromagnetic radiation whenever there is spark. The old time spark transmitters were excellent sources of man-made noise since they generated a rich noise frequency spectrum. Electric motors and fluorescent lights are also sources of man-made noise. In fact, closing or opening a switch to an electrical circuit will result in a spark that will have a frequency spectrum which interferes with the radio frequency spectrum.

Man-made noise is very difficult to predict on any basis other than statistical methods. As an example of man-made noise let us look at electrical noise in motors. Electrical noise can be of two types, that due to sparks, or an electromagnetic field.

Whenever an electrical spark occurs, an electromagnetic wave will be transmitted from the spark origin. Many frequencies are present in this wave and they can interfere with radio and television signals as well as other information systems such as computers. With the type of electric motor containing brushes, we know that a make and break action occurs with the brushes and commutator segments. Whenever this occurs it causes sparks. The motor acts as a radiator of electromagnetic waves. You can prove this to yourself by connecting a small electric motor that has brushes to a battery. Hold the motor near an AM radio and you will hear noise coming from the speaker. This noise has been produced by the electric motor in one fashion or another. It is difficult to say exactly how the noise made it to the speaker since some of the noise may have reached the loudspeaker by brute force or by frequency conversion. In brute force, noise is received by the audio amplifier or detector directly without a frequency conversion. The noise signal is so strong that it is able to take over the detector circuit or audio amplifier. The lower frequency components in the noise are then heard in the loudspeaker. With frequency conversion, high frequency components in the noise are converted downward by the frequency conversion circuits to frequencies that are in the bandpass of the intermediate frequency amplifiers. When the i-f signal is demodulated some of the components in the output are related to the spark and are heard as noise.

Other types of electric motors radiate electromagnetic waves by producing strong field patterns that can be coupled into various systems such as radios or televisions. Again either a brute force action or frequency conversion allows the system to receive the signal.

Circuit Noise

This type of noise may be caused by various processes. However, the two main processes are the thermal and shot effect.

Thermal noise is related to temperature. The thermal noise power varies proportionally with bandwidth. Thermal noise is due to electrons moving about in resistive materials due to thermal agitation in a random manner. These electrons, when moving, constitute a current flow and develop an overall voltage drop in the resistive material.

Shot noise is due to a random variation in a dc level which is present in some active devices. When electrons flow from a cathode to a plate in a vacuum tube, a current flow exists. The reason is that electrons constitute charge and it takes a certain amount of time for these electrons to move from the cathode to the plate. Since current is defined as charge per time, we can say that a current is flowing. This current is related to the average velocity of the electrons and will be called the average current. However, a few electrons will have velocities that differ from that of the average velocity electrons. These electrons will set up a random variation about the average level. If we were to listen to this variation, it would sound like lead shot hitting a concrete wall. Thus, we have the term shot noise.

It is to be noted that thermal noise is generated in active or passive components having a resistive component while shot noise is generated only in active components. Of course, a shot noise current could flow through a passive component, but it must have had its origin in an active component.

NOISE TERMINOLOGY

The main part of this book will deal with the subject of circuit noise. However for the remainder of this chapter, let us deal with electrical noise in general.

Spectral Density

Let us now consider a noise process as shown in Fig. 1-7. If

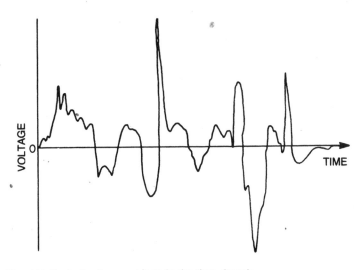

Fig. 1-7. Typical noise waveform in the time domain.

we were to examine the frequency makeup of this signal it could be shown to consist of a group of frequency components. These frequency components may differ in amplitude. The frequency makeup of this noise signal constitutes the frequency spectrum. The height or magnitude of this spectrum is called the spectral density. If the spectrum has a flat spectral density such as shown in Fig. 1-8, we say that the noise process is "white." The reason is that white light is said to be made up of all colors and white noise is said to have components at all frequencies. For the most part, white noise is constant in its spectral density. It can be shown that at extremely high frequencies white noise spectral density falls off.

Sometimes we use the term bandlimited white noise. This simply means that the white noise process we are dealing with is constant over a particular bandwidth in regards to its spectral density.

Spectral density can be related to power or voltage. Usually when we discuss spectral density we are talking about the relative noise power density. The unit of the density is given in volts squared per hertz.

In many cases, the spectral density falls off at the rate of 1/f which means that the spectral density is inversely proportional to frequency. In this case the noise process is made up of signals whose frequency components have their greatest magnitude at the lower frequencies. We know that red light is made of low frequency components. The noise has been labeled as "pink" when its spectrum falls off as 1/f. Pink noise has been studied at frequencies as low as 10^{-9} hertz (Hz) which is about one cycle in 30 years.

The spectral density for pink noise can be written as K/f, where K is a constant at 1 Hz.

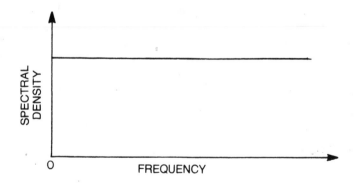

Fig. 1-8. Spectral density of white noise.

White and Pink Noise Power

White noise power is directly proportioned to bandwidth. As bandwidth increases, white noise power increases. A doubling of bandwidth increases noise power by a factor of two and increases noise voltage by the square root of two.

If you are interested in finding the square of the root-mean-square (rms) voltage in a particular white noise process, you simply multiply the spectral density by the bandwidth of interest

Example 1-1

Suppose the spectral density is 5×10^{-5} volts squared per hertz.

If the bandwidth of interest is 100 Hz then the noise voltage squared is:

(spectal density) × (bandwidth) $= 5 \times 10^{-5} \times 10^2 = 5 \times 10^{-3}$ volts^2.

We will show later that the answer we have obtained is the square of the rms noise voltage. In other words the rms noise voltage would be:

$$E_n = (5 \times 10^{-3})^{1/2} \text{ volts}$$
$$= (50 \times 10^{-4})^{1/2}$$
$$= 7.07 \times 10^{-2} = 70.7 \ mV.$$

Let us now examine a similar problem dealing with pink noise.

As a problem suppose you were to determine the noise squared voltage of a process having a spectral density which varies as $G p (f) = K/f$. Refer to Fig. 1-9.

How would you solve this problem? You could not simply multiply the spectral density by the bandwidth of interest. Instead you would have to use integration to find the area under the spectral density curve. The procedure would be as follows:

1. The mean square noise voltage is

Equation 1-3

$$E_n^2 = \int_{f_1}^{f_2} Gp(f) \, df$$

where $Gp (f)$ = spectral density and f_1, f_2 are the lower and upper frequency limits of interest.

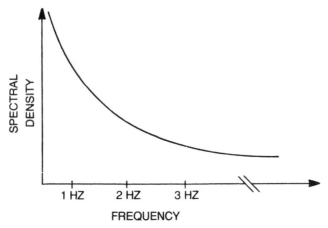

Fig. 1-9. Spectral density of pink noise.

2. Since the spectral density for pink noise is given as K/f, we can substitute this value into Equation 1-3 and obtain

Equation 1-4

$$E_n{}^2 = \int_{f_1}^{f_2} K/f \, df$$

3. Solving Equation 1-4 we obtain,

$$E_n{}^2 = \int_{f_1}^{f_2} K \, df/f = K \, (lnf) \Big|_{f_1 t}^{f_2}$$

4. Placing the limits in the equation the result is

Equation 1-5

$$E_n{}^2 = K \, (lnf_2 - lnf_1)$$

5. By a law of logarithms,

Equation 1-6

$$E_n{}^2 = K \, ln \frac{f_2}{f_1}$$

As can be seen from Equation 1-6 the square of the rms noise voltage depends on the natural logarithm of the ratio of the extreme frequencies of the bandwidth. Thus the square of the noise voltage depends on a constant K as well as the ratio of f_2 to f_1.

Example 1-2

Given a spectral density of 5×10^{-5} V^2/Hz at 1 Hz. If the frequency range of interest is from 100 to 200 Hz what is the value of E_n^2?

$$E_n^2 = K \ln f_2/f_1 = 5 \times 10^{-5} \ln (200/100)$$
$$= 5 \times 10^{-5} \ln 2$$
$$= 5 \times 10^{-5} (0.69)$$
$$= 3.45 \times 10^{-5} \text{ volts}^2.$$

E_n would then be $\sqrt{3.45 \times 10^{-5}}$

$$= \sqrt{34.5 \times 10^{-6}}$$
$$= 5.8 \times 10^{-3} \text{ V.}$$
$$= 5.8 \text{ mV.}$$

Example 1-3

What would the answer have been if the value of f_2 was 400 Hz and f_1 was 200 Hz?

Because $f_2/f_1 = 2$, the answer would have been the same.

Pin noise has the same mean square noise voltage as long as the frequency ratios are the same. The noise voltage depends on the ratio of f_2 to f_1 and not like white noise, which depends on the difference between f_2 and f_1.

PROBABILITY AND STATISTICS FOR NOISE

No book dealing with the subject of noise would be complete without some information on probability and statistics and how a noise process relates to such subjects.

Webster's Dictionary defines probability as "that which is likely to happen; the state of being likely to happen." In mathematics we could define probability as the study of the average behavior of events which leads to the frequency of occurrence of the various possible outcomes.

What is probability? Let us approach an understanding of this subject by thinking about the subject of coin tossing.

Suppose we perform a coin tossing experiment in which we have two possible outcomes, namely heads or tails. Let X equal heads and Y equal tails. If w_x are the number of occurrences of X and w_y are the number of occurrences of Y, the frequency of occurrences in n tosses could be given as w_x/n or w_y/n. Should the number of tosses be very large, then the ratio w_x/n and w_y/n will become some average value. The limit that these ratios approach as the number of tosses approaches infinity is called their probability of occurrence. Namely,

Equation 1-7

$$P(X) = \lim_{n \to \infty} w_x/n$$

or

Equation 1-8

$$P(Y) = \lim_{n \to \infty} w_y/n$$

$P(X)$ is the probability of outcome for x and $P(Y)$ is the probability of outcome for y.

Should a particular outcome always occur such as w_x always occurs in n tosses, then the probability is one. Should w_x never occur its probability is zero. We can see that the limits of probability lies between zero and one.

In summary,

Equation 1-9

$$0 \leq P(X) \leq 1 \text{ and } 0 \leq P(Y) \leq 1$$
$$P(X) + P(Y) = 1$$

Functions of Probability

When observing a process such as random noise, the noise voltage could take on various values such as V_1, V_2, ... V_n and the probability for each is given as $P(V_1)$, $P(V_2)$ ... $P(V_1)$. A plot of these gives us a probability function like is shown in Fig. 1-10. To obtain the probability that the voltage is between a value of V_1 and V_2, it is necessary to find $P(V_1 < V < V_2)$ which leads to the concept of the probability distribution function where

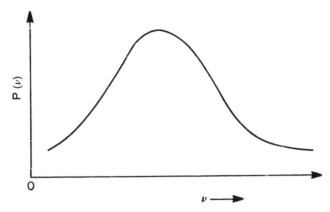

Fig. 1-10. A probability distribution function.

Equation 1-10

$$P\,(V_1 \le V - V_2) = \int_{V_1}^{V_2} P\,(x)\,dx$$

where $P(x)$ is the probability distribution function.

Mean Value of a Variable

For a discrete set of values such as V_1, V_2, etc., of a random variable V with occurrences n_1, n_2, etc., the mean value of the variable is given by Equation 1-11.

Equation 1-11

$$\overline{V} = \frac{V_1 n_1 + V_2 n_2 + \ldots V_k n_k}{n}$$

where $n = n_1 + n_2 + \ldots n_k$

When N is very large, n_1/n, n_2/n are probabilities $P(V_1)$, $P(V_2)$ respectively. The result is Equation 1-12.

Thus,

$$\overline{V} = V_1\,P(V_1) + V_2\,P\,(V_2) + \ldots$$

Equation 1-12

$$V = \sum_{i\,=\,1}^{i\,=\,k} V_i\,P\,(V_i)$$

17

Variance

This is a set of discrete values such as V_1, V_2, etc. whose mean value is V. Their deviations with respect to the mean are $V_1 - V$, $V_2 - V$, etc. If you were to square each one of these deviations the results would always be positive. The average of the sum of the square of these deviations is called the variance.

In the subject of noise, variance corresponds to the mean square noise voltage and/or mean square noise current. In formula form,

Equation 1-13

$$\sigma^2 = \frac{(V_1 - \overline{V})^2 + (V_2 - \overline{V})^2 + - - - (V_n - \overline{V})^2}{n}$$

Standard Deviation

The standard deviation is the quantity which gives the spread of values about the mean value. Standard deviation is the square root of the variance. The smaller the standard deviation, the closer are the values to the mean. In noise, standard deviation represents the root mean square voltage or current.

If we have two noise processes of standard deviations V_1 and V_2 the variance of their sum can be shown as

Equation 1-14

$$\sigma^2 = V_1^2 + V_2^2$$

When very large numbers of statistical results are analyzed some order can be obtained from the random behavior of the values. One of the most important distributions of statistical results is called the normal or Gaussian distribution.

Normal or Gaussian Distribution

For a random variable X, the normal probability density function Pd(x) is given by the formula Equation 1-15.

Equation 1-15

$$Pd(x) = \frac{1}{\sigma\sqrt{2\pi}} \cdot e^{-(X-M)^2/2\sigma^2}$$

The appearance is as in Fig. 1-12.

The most probable value occurs when $X = M$, where M is the mean value and σ is the standard deviation. The curve for proba-

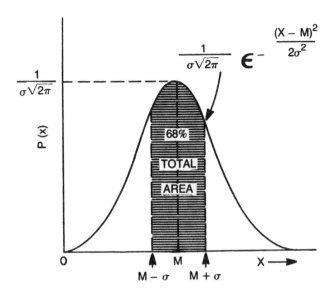

Fig. 1-11. The Gaussian distribution function.

bility function is the well-known bell shape with symmetry about X = M. This means that there is an equal probability of a random value falling below or above X = M. We can see the occurrence in Fig. 1-11. In a noise process M would be similar to an average level and σ the rms value of the noise. The quantity P (X) is the spot probability of a certain occurrence.

The area under the curve between M + σ and M − σ is 68 percent of the entire area. It is much more likely that small deviations from the mean will occur than large deviations. It is extremely unlikely that deviations larger than four standard deviations will occur.

SUMMARY

1. Noise is normally thought of as a randomly varying signal which interferes with the transmission and reception of information through a network.

2. Electrical noise can be broken down into a group of categories. They are:

A. Erratic disturbances which may be due to atmospheric or space noise.

    B. Man-made due to electrical machines and ignition systems.

    C. Circuit noise which is caused by resistors and active components.

    3. Two important classifications of noise are white and pink. White noise has a spectral density that for all practical purposes is considered flat. Pink noise has a spectral density that falls off inversely to frequency.

    4. The root mean square value of a noise process is the same as the standard deviation.

    5. The mean square voltage or current is the same as the variance.

    6. One very common probability distribution function is the Gaussian curve.

PROBLEMS AND QUESTIONS

    1. List the three main categories for noise and discuss each one.

    2. Explain why lightning interferes with the transmission of radio and television signals.

    3. Write a report on radio astronomy discussing such things as the earth's optical and electrical window.

    4. It is noted that a small dc motor causes interference on a television when it is held near the set. Why does the interference occur and when during the operating cycle of the motor does the interference occur?

    5. What are the differences between shot and thermal noise?

    6. What is the significance of describing a noise process as white or pink?

    7. For white noise, what will happen to noise power if bandwidth doubles? What would happen to a pink noise process for a doubling of bandwidth?

    8. The spectral density of a white noise process is 7×10^{-8} volts ^{2}/Hz. If the bandwidth of interest is 100 kHz what is the mean squared output voltage?

    9. Suppose that the spectral density of a pink noise is 8×10^{-9} volts ^{2}/Hz at 1 Hz. What is the mean squared output voltage in a frequency range of interest from 600 to 2400 Hz?

10. Given the following set of data:

100.7
111.9
105.6
 99.4
 95.6

Compute the mean value of the variable, the variance, and the rms value (standard deviation).

11. Explain the Gaussian curve and how it relates to noise.

12. For the model of the lightning induced voltage in Fig. 1-5, write equations that describe the waveform. Assume the rise is a ramp function and the fall is exponential.

13. Of what importance is the northern and southern lights in regards to noise? What is their cause?

14. How would you define cosmic and solar noise?

15. Estimate the power dissipated during a typical lightning stroke.

16. A sphere has a charge of $16\mu C$ present on it. Estimate the voltage at a point 12 cm from the sphere.

Chapter 2

Basic Signals and Systems For Noise

$\mathbf{A}$ SIGNAL IS THE FORM ENERGY TAKES TO PROPAGATE through a circuit. Signals are represented by math models to be able to decide which signal is the most useful in describing information and to analyze various circuits according to their excitation.

CLASSIFICATION OF SIGNALS

A periodic signal repeats itself exactly after a fixed length of time known as the period. A signal may be written as a function of time as V(t). If there is a number T such that $V(t) = V(t + T)$ for every positive value t may assume, we say we have a periodic signal. Refer to Fig. 2-1. As you see the waveform repeats itself every .1 sec. Thus the waveform has the same value at .4 sec as it does at .5 sec. The smallest value of time that satisfies the equation for T is called the period. The reciprocal of this period is called the fundamental frequency.

Equation 2-1

$$\text{Fundamental frequency} = \frac{1}{\text{Period}}$$

or

$$f = \frac{1}{T}$$

Examples of typical periodic signals are, sinusoids, sawtooths, and squarewave waveforms.

Non Periodic Signal

A non-periodic signal is one which no value of T satisfies the expression $V(t) = V(t + T)$. Examples of non-periodic signal are speech, transients and random signals.

Random Signals

A random signal is one in which there is uncertainty before it actually occurs. In other words, we don't know what the signal is going to look like or what values it will assume before occurrence.

Examples are signals due to atmospheric disturbances, man-made interference, and circuit noise.

SIGNAL PARAMETERS

A waveform may have an average level associated with it. The average level is the area under the waveform in study divided by the period. The average level can be either in volts or amperes.

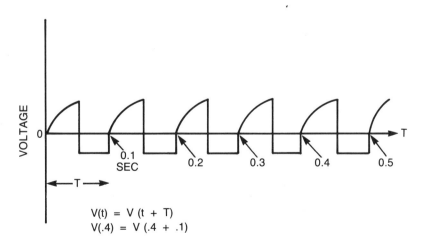

Fig. 2-1. Example of a periodic waveform.

The formula is:

Equation 2-2

$$Eav = \frac{\text{Area under waveform over one period}}{\text{period}}$$

For a simple waveform we can easily find the average level.

Example 2-1

Find the average level of the waveform shown, in Fig. 2-2A using Equation 2-1.

1. Area under waveform: (20 Volts) (1 Sec.)
2. Period = 2 Sec.
3. Average level = $\frac{(20\,V)(1)}{2}$ = 10 Volts.

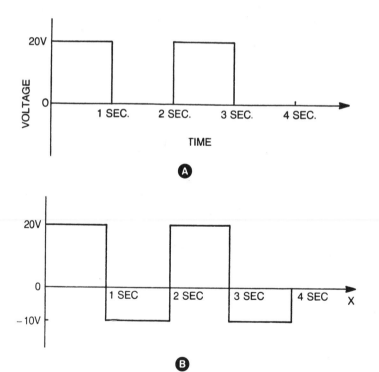

Fig. 2-2(A). Waveform for Example 2-1. (B) Waveform for Example 2-2.

Example 2-2
Find the average level of the waveform shown in Fig. 2-2B using Equation 2-1.

1. Area under waveform: (20 Volts) (1 Sec) + (– 10 Volts) (1 Sec)

$$= 20 \text{ volt-sec} - 10 \text{ volt-sec}$$
$$= 10 \text{ volt-sec}$$

2. Period = 2 sec.

3. Average level = $\dfrac{10 \text{ volt-sec}}{2 \text{ sec}}$ = 5 Volts.

For more complex waveforms we must go to calculus. The formula is given by Equation 2-2.

Equation 2-2

$$Eav = \frac{\int_0^T e\,(t)\,dt}{T}$$

where e(t) is the signal in question and T is the period.

Example 2-3
Find the average level of the signal shown in Fig. 2-3 using Equation 2-2.

1. $V(t)$ = 4t for 0 < t < 0.1 sec.
2. T = .1 sec.

3.

$$Eav = \frac{\int_0^{.1} 4t\,dt}{0.1} = \left[\frac{4t^2}{2(.1)}\right]_0^{0.1}$$

$$= \left[20t^2\right]_0^{0.1}$$

$$= 20(0.1)^2 - 20(0)^2$$
$$= 0.2 \text{ volt}$$

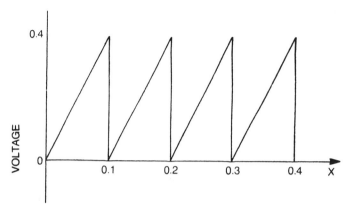

Fig. 2-3. Waveform for Examples 2-3 and 2-5.

Effective Value

The effective value of an alternating waveform is that value, whether voltage or current, which cause the same power dissipation in a given resistance as direct voltage or current of the same numerical value. The effective value is also called the root mean square value since it is the square root of the average of the squares of the instantaneous values of the waveform over one period.

The formula is:

Equation 2-3

$$E_{rms} = \sqrt{\frac{\text{Area under square of waveform}}{\text{Period}}}$$

Example 2-4

Find the rms value of the waveform in Fig. 2-4A.

1. Area under square of waveform: $(5 \text{ volts})^2 \times 1 \text{ sec.} = 25$ volts^2 sec.

2. Period $= 2$ sec.

3. $Erms = \sqrt{25/2} = 3.535$ volts.

In other words, a periodic waveform consisting of five volt pulses lasting one sec and repeating in two sec applied across a resistance, has the equivalent heating effect as a 3.535 volt dc voltage across the same resistance.

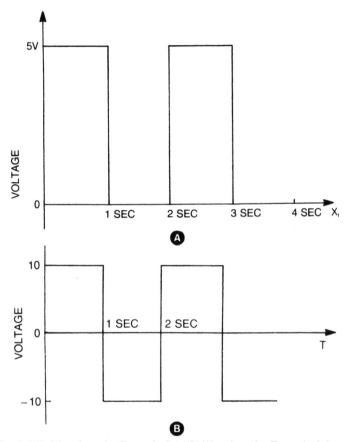

Fig. 2-4(A). Waveform for Example 2-4. (B) Waveform for Example 2-8.

In general, the effective or rms voltage value of a signal may be obtained by using the formula:

Equation 2-4

$$Erms = \sqrt{\frac{\int_0^T (V(t))^2 \, dt}{T}}$$

Example 2-5

Find the rms voltage value for the waveform shown in Fig. 2-3 using Equation 2-4.

1. $V\,(t)\ =\ 4t$ for $0<t<0.1$ sec.
2. $[V(t)]^2\ =\ 16t^2$ for $0<t<0.1$ sec.

3. $Erms\ =\ \sqrt{\dfrac{\displaystyle\int_0^{0.1} 16\ t^2\ dt}{0.1}}$

$$=\ \sqrt{\dfrac{\left.\dfrac{16\ t^3}{3}\right|_0^1}{.1}}$$

$$=\ \sqrt{\dfrac{5.33\,(0.1)^3\ -\ 5.33\,(0)^3}{0.1}}$$

$$=\ \sqrt{5.33\,(.01)}$$

$$=\ \sqrt{.0533}$$

$$=\ 0.23\ \text{volts.}$$

Form Factor

The form factor of a waveform is the ratio of the root mean square value to the average value. The form factor gives information about the shape forming characteristics of the waveform.

Equation 2-5

$$FF\ =\ E_{rms}/E_{av.}$$

Example 2-6

A sinewave has an rms voltage of 0.707 times its peak value. Also, its full wave rectified value is 0.636 times its peak value. What is the form factor?

Equation 2-6

$$FF\ =\ \frac{0.707\,(E_{peak})}{0.636\,(E_{peak})}\ =\ 1.11$$

Crest Factor

The crest factor of a waveform is the ratio of the peak value to the rms value. The crest factor gives us information regarding the peak amplitude in relation to the rms value. Especially in noise, the crest factor may be quite large since a large voltage may occur occasionally. But it is more normal to find small voltage values. The rms voltage may then be quite small giving large crest factors.

Example 2-7

A waveform has a peak value of 20 volts. The rms value of the waveform is 15 volts. Compute the crest factor.

Crest Factor = 20 V/15 V = 1.33.

Example 2-8

Compute the crest factor and form factor for the waveform shown in Fig. 2-4B. Use Equations 2-5 and 2-6 respectively.

1. Full wave average

$$= \frac{10 \text{ volts (1 sec)} + 10 \text{ volts (1 sec.)}}{2 \text{ sec.}} = 10 \text{ volts}$$

2. Peak value (by inspection) = 10 volts.

3. Root mean square value

$$= \sqrt{\frac{(10 \text{ volts})^2 (1 \text{ sec}) + (-10 \text{ volts})^2 (1 \text{ sec})}{2 \text{ sec.}}}$$

$$= \sqrt{100 \text{ volts}^2} = 10 \text{ volts.}$$

4. Form factor $= \dfrac{\text{rms}}{\text{average}} = \dfrac{10 \text{ volts}}{10 \text{ volts}} = 1.$

5. Crest factor $= \dfrac{\text{peak}}{\text{rms}} = \dfrac{10 \text{ volts}}{10 \text{ volts}} = 1.$

Frequency Make Up of Signals

A periodic signal having a fundamental period T can be expressed as a sum of an average level and a group of sinusoidal waveforms. The sum gives us what we call the Fourier series.

We may write the series as follows:

Equation 2-7

$$f(t) = A_0 + \sum_{n=1}^{\infty} \frac{An(\cos 2\pi nt)}{T} + \sum_{n=1}^{\infty} \frac{Bn(\sin 2\pi nt)}{T}$$

In words, the formula states that the Fourier series of a function f(t) is the sum of an average level A_0, plus the summation of a group of cosine waves with amplitudes An and the summation of a group of sinewaves having amplitudes Bn.

If we examine these summations we see terms like An cos $2\pi nt/T$ or Bn sin $2\pi nt/T$. The "n" in the amplitude indicates to which particular signal we are referring. Symbols A_1 and B_1 indicate the fundamental amplitudes. The fundamental frequency is $2\pi n/T$ where n = 1 or $2\pi/T$. A_2 and B_2 indicate the second harmonic amplitudes. The second harmonic frequency is 2 $\pi n/T$ where n = 2 or $2\pi(2)/T$. Likewise for n = 3 we have the third harmonic with amplitudes A_3 and B_3 and frequencies $2\pi(3)/T$.

To calculate Ao, An, and Bn, we use the following formulas:

Equation 2-8

$$A_o = \frac{1}{T} \int_{-\frac{T}{2}}^{\frac{T}{2}} f(t)\, dt$$

Equation 2-9

$$A_n = \frac{2}{T} \int_{-\frac{T}{2}}^{\frac{T}{2}} \frac{f(t)\,(\cos 2\pi nt)dt}{T}$$

Equation 2-10

$$B_n = \frac{2}{T} \int_{-\frac{T}{2}}^{\frac{T}{2}} \frac{f(t)\,(\sin 2\pi nt)dt}{T}$$

We note that formulas for A_o, A_n, and B_n have their limits as $-T/2$ and $T/2$ where T is the period. The limits could have been just as well as O and T.

Example 2-9

Determine the Fourier series for the waveform shown in Fig. 2-5.

1. The function can be written as

$f(t)$ = V for $-r/2 < t < r/2$

= 0 elsewhere.

2. The average level will be given as:

$$Ao = \frac{1}{T} \int_{-\frac{T}{2}}^{\frac{T}{2}} f(t)\, dt$$

Since the waveform exists only between $-r/2$ and $r/2$, then

$$Ao = \frac{1}{T} \int_{-\frac{r}{2}}^{\frac{r}{2}} V\, dt = \frac{1}{T}\, (V)t \, \Big|_{-\frac{r}{2}}^{\frac{r}{2}}$$

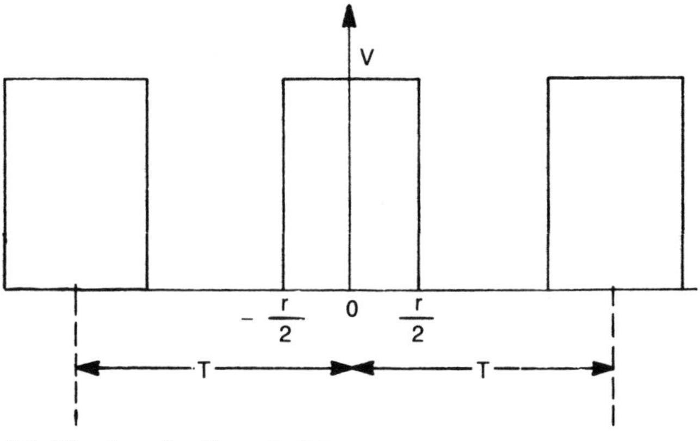

Fig. 2-5. Waveform for Example 2-9.

$$= \frac{V}{T}\left[\frac{r}{2} - \frac{(-r)}{2}\right] = \frac{V(r)}{T}$$

$$A_0 = V(r)/T.$$

3. The amplitudes of the cosine waves can be given by

$$A_n = \frac{2}{T}\int_{-\frac{T}{2}}^{\frac{T}{2}} f(t)\,(\cos 2\pi nt/T)\,dt$$

$$= \frac{2}{T}\int_{-\frac{r}{2}}^{\frac{r}{2}} V\,(\cos 2\pi nt/T)\,dt$$

$$= \frac{(V)2}{T}\int_{-\frac{r}{2}}^{\frac{r}{2}} (\cos 2\pi nt/T)\,dt$$

$$= \frac{2}{T}\,(V\left[\frac{\sin 2\pi nt/T}{\frac{2\pi n}{T}}\right]_{-\frac{r}{2}}^{\frac{r}{2}}$$

$$A_n = 2/T\,(V)\left[\frac{\sin 2\pi n/T\,(r/2) - \sin 2\pi n/T\,(-r/2)}{2\pi n/T}\right]$$

$$= \frac{2/T\,(V)\sin \pi nr/T - \sin(-\pi nr/T)}{\frac{2\pi n/T}{T}}$$

$$= \frac{V\,[\sin(\pi nr/T) - \sin(-\pi nr/T)]}{\pi n}$$

Since the sine of an angle follows a behavior $\sin x = -\sin(-x)$, then

$$\sin(\pi nr/T) = -\sin(-\pi nr/T).$$

Thus, $A_n = \dfrac{V(2 \sin \pi n r/T)}{\pi n}$

$A_n = 2V/\pi(n)$ (sin $\pi n r/T$.)

Evaluation of A_n is as follows: Since $r/T = 1/2$,

A_1 = $2V/\pi(1) \sin \pi (1) (1/2)$

    = $2V/\pi \sin \pi/2 = 2V/\pi(1) = 2V/\pi$

A_2 = $2V/2\pi \sin \pi(2) (1/2)$

    = $2V/2\pi \sin \pi = V(0) = 0$

A_3 = $2V/\pi(3) \sin \pi(3) (1/2)$

    = $2V/3\pi \sin 3\pi/2 = 2V/3\pi(-1) = -2V/3\pi$

A_4 = $2V/\pi(4) \sin \pi (4) (1/2)$

    = $2V/4\pi \sin 2 \pi = V/2\pi(0) = 0$

A_5 = $2V/\pi(5) \sin \pi (5) (1/2)$

    = $2V/5\pi \sin 5\pi/2 = 2V/5(+1) = +2V/5\pi$

It is obvious by now that the even values for n give A_n values that are zero. Likewise, odd values for n give A_n values.

Figures 2-6 and 2-7 shows these values.

We notice that $n = 1, 5$, give positive values for A_n, while n = 3, 7, give negative values.

A negative value simply means that the cosine waveform is starting from a negative maximum.

4. The amplitudes of the sinewaves can be given by Bn

N (ODD)	A_n
1	$\dfrac{2V}{\pi}$
3	$-\dfrac{2V}{3\pi}$
5	$\dfrac{2V}{5\pi}$
7	$-\dfrac{2V}{7\pi}$

Fig. 2-6. A_n values for odd n values in Example 2-9.

N (EVEN)	A_n
0	$\dfrac{V}{2}$
2	0
4	0
6	0
8	0

Fig. 2-7. A_n values for even n values in Example 2-9.

$$Bn = \frac{2}{T} \int_{-\frac{r}{2}}^{\frac{r}{2}} f(t)\,(\sin 2\pi nt/T)\,dT$$

$$= \frac{2}{T} \int_{-\frac{r}{2}}^{\frac{r}{2}} (V)\,(\sin 2\pi nt/T)\,dt$$

$$= 2(V) \int_{-\frac{r}{2}}^{\frac{r}{2}} \sin 2\pi nt/T\,dt$$

$$= 2\,(V)/T \left[\frac{-\cos 2\pi nt/T}{2\,\pi n/T} \right]_{\frac{-r}{2}}^{\frac{r}{2}}$$

$$= 2(V)/T \left[\frac{-\cos 2\pi n/T\,(r/2) - (-\cos 2\pi n/T\,(-r/2)}{\dfrac{2\pi n}{T}} \right]$$

$$= V\,\frac{-\cos \pi nr/T + \cos(-\pi nr/T)}{\pi n.}$$

Since the cosine of an angle follows a behavior $\cos x = \cos(-x)$ then

$$\cos\,(\pi nr/T) \;=\; \cos\,(-\pi nr/T.)$$

Thus,

$$Bn \;=\; V\,\frac{-\cos\,(\pi nr/T)\;+\;\cos\,(\pi nr/T)}{\pi n.}$$

$$Bn \;=\; 0$$

Thus we can say that Bn is zero for all values of n.

The Fourier series can now be stated for this example. It is:

$$n = \infty$$
$$V\,(t) \;=\; V/2 \;+\; (\Sigma\,_{n=1}\,2V/\pi n\,\cos\,2\pi nt/T.$$

The series when evaluated for the first five values of n gives

$$V\,(t) \;=\; V/2 \;+\; 2V/\pi\,\cos\,2\pi t/T \;-\; 2V/3\pi\,\cos\,2\pi(3)t/T$$
$$+\; 2V/5\pi\,\cos\,2\pi(5)t/T \;+\; \dots$$

This may be simplified to:

$$V\,(t) \;=\; V/2 \;+\; 2V/\pi\,\cos\,2\pi t/T \;-\; 2V/3\pi\,\cos\,6\pi t/T$$
$$+\; 2V/5\pi\,\cos\,10\pi t/T.$$

Figure 2-8 shows the frequency spectrum for the waveform.

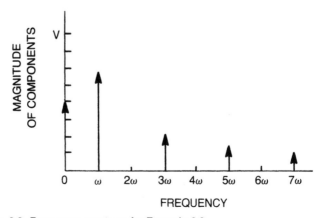

Fig. 2-8. Frequency spectrum for Example 2-9.

Example 2-10

Determine the Fourier series for the waveform shown in Fig. 2-9.

1. The function can be written as

$$f(t) = V \text{ for } 0 < t < T/2$$
$$= -V \text{ for } T/2 < t < T.$$

2. The average level will be given as:

$$A_o = \frac{1}{T} \int_{-\frac{T}{2}}^{\frac{T}{2}} f(t) \, dt$$

$$= \frac{1}{T} \int_{0}^{\frac{T}{2}} V \, dt + \frac{1}{T} \int_{\frac{T}{2}}^{T} (-V) \, dt$$

$$= 1/T \, [V \times (T/2 - o)] + 1/T \, [(-V) \, (T - \frac{T}{2})]$$

$$= \frac{V}{2} - \frac{V}{T} \frac{(T)}{2} = \frac{V}{2} - \frac{V}{2} = 0.$$

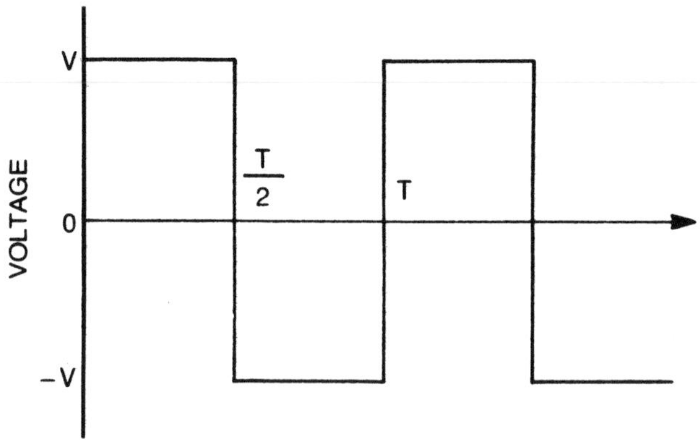

Fig. 2-9. Waveform for Example 2-10.

The average level is thus zero for this waveform.

3. The amplitudes of the cosine waves can be given by

$$An = \frac{2}{T} \int_{0}^{\frac{T}{2}} V (\cos 2\pi nt)\, dt + \frac{2}{T} \int_{\frac{T}{2}}^{T} -V (\cos 2\pi nt)\, dt$$

$$= 2V/T \left[\frac{\sin \dfrac{2\pi nt}{T}}{\dfrac{2\pi n}{T}} \right]_{0}^{T/2} - 2V/T \left[\frac{\sin \dfrac{2\pi nt}{T}}{\dfrac{2\pi n}{T}} \right]_{+ T/2}^{T}$$

$$= 2V/T \left[\frac{\sin \dfrac{2\pi n}{T}\,(T) - \sin \dfrac{2\pi n}{T}\,(0)}{2\pi n/T} \right]$$

$$- 2V/T \left[\frac{\sin \dfrac{2\pi n}{T}\,(T) - \sin \dfrac{2\pi n}{T}\,\dfrac{T}{2}}{2\pi n/T} \right]$$

$$= \frac{V}{\pi(n)} \Big[\sin \pi n - \sin 0 \Big] - \frac{V}{\pi(n)} \Big[+ \sin 2\pi - \sin \pi n \Big]$$

$$= V/\pi(n)\,(0 - 0) - V/\pi n\,(0 - 0)$$

$$= 0.$$

Thus all An values are zero.

4. The amplitudes of the sine waves can be given by

$$Bn = \frac{2}{T} \int_{0}^{\frac{I}{2}} V (\sin 2\pi nt)\, \frac{dt}{T} + \frac{2}{T} \int_{\frac{T}{2}}^{T} (-V)\, \frac{(\sin 2\pi nt)\, dt}{T}$$

$$Bn = 2V/T \left[\frac{-\cos 2\pi nt/T}{2\pi n/T} \right]_{0}^{T/2} - 2V/T \left[\frac{-\cos 2\pi nt/T}{\dfrac{2\pi n}{T}} \right]_{T/2}^{T}$$

37

$$= V/\pi n \quad [-\cos 2\pi n/T \, (T/2) - (-\cos 2\pi n/T \, (0))]$$
$$- V/\pi n \quad [-\cos 2\pi n/T \, (T) - (-\cos 2\pi n/T \, (T/2))]$$
$$= V/\pi n \quad [-\cos \pi n + \cos \pi \, 0]$$
$$- V/\pi n \mathrm{T} \quad [-\cos 2\pi n + \cos \pi n]$$
$$= V/\pi n \quad [-\cos \pi n + 1 + \cos 2\pi n - \cos \pi n]$$
$$= V/\pi n \quad [-\cos \pi n + 1 + 1 - \cos \pi n]$$
$$= 2V/n \quad [1 - \cos \pi n.]$$

Evaluation of Bn is as follows:

$$B_1 = 2V/\pi \, [1 - \cos \pi(1)] = 2V/\pi \, [1 - (-1)] = 4V/\pi$$
$$B_2 = 2V/2\pi \, [1 - \cos \pi(2)] = 2V/2\pi \, [(1 - 1)] = 0$$
$$B_3 = 2V/3\pi \, [1 - \cos \pi(3)] = 2V/3\pi \, [1 - (-1)] = 4V/3\pi$$
$$B_4 = 2V/4\pi \, [1 - \cos \pi \, (4)] = 2V/4\pi \, [1 - 1] = 0$$
$$B_5 = 2V/5\pi \, [1 - \cos \pi \, (5)] = 2V/5\pi \, [1 - (-1)] = 4V/5\pi.$$

If we write the Fourier series we will have:

$$V_{(t)} = 4V/\pi \, \sin 2\pi t/T + \frac{4V}{3\pi t} \sin 6\pi t/T + 4V/5\pi \sin 10 \, \pi t/T$$
$$+ - - - -$$

Figure 2-10 shows the frequency spectrum for the waveform. If we were to add graphically all of the components in the Fou-

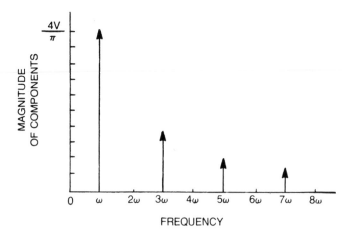

Fig. 2-10. Frequency spectrum for Example 2-10.

rier series we would obtain a signal that would look similar to the original waveform. The more components used, the closer the approximation will be to the original.

SYSTEMS

An electrical system can be thought of as a group of components connected in a particular manner to perform an electrical function. The system has an input and output terminal. A signal is applied to the input and the system then performs some operation on the signal. At the output a new signal appears. The system may simply amplify, integrate, differentiate, square or some other basic operation. Or the system may perform a complex operation.

Regardless of the type of system we are dealing with, there is always some filtering action inherent in the system. The filtering action may be brought on by internal components or distributed parameters such as shunt inductance and capacitance.

Filters

To understand how noise is affected when passing through a system, we must understand the concept of the electric filter.

We could define an electric filter as a network that transforms an input signal in some specified manner to yield a desired output signal. The filter may be looked at as a frequency-selective device which passes signals of certain frequencies and blocks or attenuates signals of other frequencies. Also, a filter will change the phase of certain signals as they pass through the filter network.

We may classify a filter in several ways. Analog filters process analog signals. These are signals which are functions of continuous-time variables. Digital filters will operate on signals that are digitized.

Also, a filter may be classified as either passive or active. Passive filters contain only passive components like resistors, capacitors and inductors. Active filters contain at least one active device, as well as passive components. Such active devices are transistors and vacuum tubes.

Types of Filters

Filters normally can be grouped into one of the following classifications:

● Low pass filters pass a group of signals with little attenuation up to the cutoff frequency. For signals with frequencies higher than the cutoff frequency, increasing attenuation occurs. Higher frequencies are attenuated.

● High pass filters give appreciable attenuation to frequencies between zero and the cutoff frequency. Above the cutoff frequency attenuation becomes smaller and the filter passes signals easier.

● Bandpass filters pass a group of frequencies with little attenuation between two cutoff frequencies, namely the upper cutoff W_u and the lower cutoff W_1. For other frequencies appreciable attenuation occurs. The difference between the two cutoff frequencies is called the bandwidth.

● A band reject filter has a frequency response that is opposite to that of a bandpass filter. Frequencies between zero and the lower cutoff W_1 and frequencies beyond the higher cutoff W_u are passed with little attenuation. Those signals whose frequencies lie between W_1 and W_u are attenuated appreciably. If the frequency range of attenuation is very narrow, we may refer to the filter as a "notch" filter.

● "All pass filters" will pass all frequencies with little attenuation. However, phase shift varies for the various input frequencies as they pass through the networks. The various filter responses are shown in Fig. 2-11.

Ideal Filter

Quite often in the study of noise we must equate an actual filter to an ideal filter to make calculations. Although an ideal filter could be shown for all of the filters mentioned above, we usually mention the ideal low pass filter. Such a filter has no attenuation up to the cutoff frequency. After the cutoff frequency the response drops to zero with an infinite slope. For all frequencies after the cutoff frequency we have infinite attenuation.

Cutoff Frequency

The cutoff frequency of a filter is that frequency where the filter response drops to 0.707 of the filter response maximum or flat response. This can be seen through Fig. 2-12.

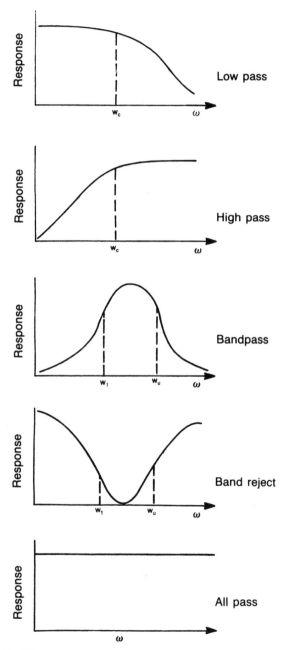

Fig. 2-11. Filter responses.

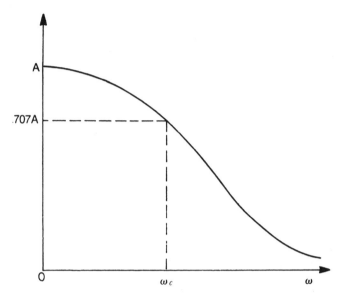

Fig. 2-12. Illustration of cutoff frequency.

Transfer Functions

The transfer function of a filter describes how the filter output voltage input voltage ratio varies in respect to frequency. To understand how to calculate a transfer function refer to Fig. 2-13.

The input voltage for the simple voltage divider is Ei. This could be written as:

$E_i = i (R + R)$

The output voltage is:

$E_o = i (R).$

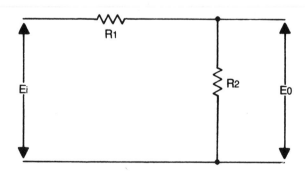

Fig. 2-13. Basic resistive voltage divider.

If we are to take the ratio of E_o to E_i we obtain

Equation 2-12

$$\frac{E_o}{E_i} = \frac{iR_2}{i(R_1 + R_2)} = \frac{R_2}{R_1 + R_2}$$

Now we can see that there is no frequency present in the transfer function. This means that regardless of the input signal frequency the output will be simply the resistive ratio $R_2/R_1 + R_2$ multiplied by the input signal.

Now let us change R_1 and R_2 to impedances Z_1 and Z_2 respectively. The ratio of E_o to E_i now becomes:

Equation 2-13

$$\frac{E_o}{E_i} = \frac{Z_2}{Z_1 + Z_2}$$

We may write E_o and E_i as $E_{o(jw)}$ and $E_{i(jw)}$ respectively indicating they are functions of frequency.

Example 2-11

Find the transfer function of the circuit shown in Fig. 2-14. First, define the various impedances.

$Z_a = jwL$

$Z_b = R$

$Z_c = 1/jwc.$

Second, determine the transfer function

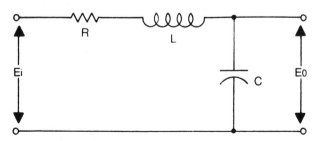

Fig. 2-14. Circuit for Example 2-11.

43

Equation 2-14

$$\frac{E_o\,(jw)}{E_i\,(jw)} = \frac{iZ_c}{i(Z_a + Z_b + Z_c)}$$

$$= \frac{Z_c}{Z_a + Z_b + Z_c}$$

$$= \frac{\dfrac{1}{jwc}}{jwL + R + 1/jwc.}$$

Multiply the function by *jwc* to simplify numerator and denominator.

Equation 2-15

$$\frac{E_o(jw)}{E_i(jw)} = \frac{1}{-w^2LC + jwC\,R + 1}$$

The magnitude of the function is given by the formula

$$\frac{E_o\,(jw)}{E_i\,(jw)} = \frac{1}{\sqrt{(1 - \omega^2LC)^2 + (\omega CR)^2}}$$

$$= \frac{1}{\sqrt{1 - 2\,\omega^2\,LC + \omega^2C^2R^2 + \omega^4L^2C^2}}.$$

If we were to place various frequencies into the magnitude, we would find that as frequency became very large that magnitude would become very small. This transfer function constitutes a low pass filter.

Example 2-12

Find the transfer function for the network shown in Fig. 2-15.
1. Write mesh equations for the circuit

$$E_i = I_1\,(R + 1/jwc_1) - I_2\,(1/jwc_1)$$

$$0 = -I_1\,(I/j\,\omega\,C_1) + I_2\,(1/j\,\omega\,C_1 + R + 1/\omega\,C_2)$$

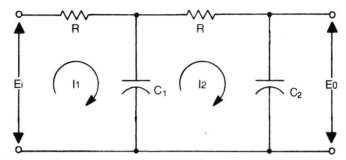

Fig. 2-15. Circuit for Example 2-12.

2. Find the determinant of the system

$$= \begin{vmatrix} R + 1/jwc1, & - 1/jwc \\ - 1/jwc, & 2/jwc + R \end{vmatrix}$$

$$= \left[R + \frac{1}{jwc} \right]\left[R + \frac{2}{jwc} \right] - \frac{(-1)}{jwc}\frac{(-1)}{jwc}$$

$$= R^2 + \frac{2R}{jwc} + \frac{R}{jwc} - \frac{2}{w^2c^2} + \frac{1}{w^2C^2}$$

Equation 2-17

$$= R^2 + \frac{3R}{jwc} - \frac{1}{w^2C^2} \cdots$$

3. Find I_2.

$$\Delta I_2 = \begin{vmatrix} R + 1/jwc, & E\,(jw) \\ - 1/jwc, & 0 \end{vmatrix}$$

$$= (R + \frac{1}{jwc})\,(0) - (-1)\,(E\,(jw))}{jwc}$$

$$= \frac{Ei}{jwc.}$$

4. Find $I_2 = \Delta I_2/\Delta = \dfrac{\dfrac{Ei}{jwc}}{R^2 + \dfrac{3R}{jwc} - \dfrac{1}{w^2\,C^2}}$

5. Compute $Eo = I_2 (1/jwc)$

$$= \frac{Ei \ (1/jwc) \ (1/jwc)}{R^2 + 3R/jwc - 1/w^2c^2}$$

$$= \frac{Ei \ (-1/w^2c^2)}{R^2 + 3R/jwc - 1/w \ c^2.}$$

6.

$$E_o/E_i = \frac{-1/w^2c^2}{R^2 - 1/w^2c^2 + 3R/jwc}$$

Multiply top and bottom by w^2c^2

$$E_o/E_i = \frac{-1}{R^2 w^2 C^2 - 1 - j3R \ w \ C}$$

Equation 2-18

$$= \frac{1}{1 - w^2 C^2 R^2 + j3 \ w \ R \ C}$$

The formula derived is the transfer function of a low pass filter.

Example 2-13

Determine the transfer function for the following circuit in Fig. 2-16.

1. Gain of circuit 1 is A.
2. Transfer function of circuit 2 is

$$= \frac{1/jwc}{1/jwc + R} = \frac{1}{j \ wCR + 1}$$

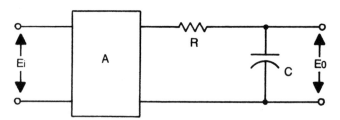

Fig. 2-16. Circuit for Example 2-13.

3. The overall transfer function is:

Equation 2-19

$$\frac{E_o}{E_i} = \frac{A}{1 + jwCR}$$

Example 2-14

For several circuits connected in cascade, and isolated from each other, the overall transfer function is the product of the individual functions.

$$E_o/E_i = \frac{1}{1 + jwCR} \times \frac{1}{1 + jwCR}$$

$$= \frac{1}{(1 + wCR)^2}$$

Equation 2-20

$$= \frac{1}{1 + j2\ wCR - w^2C^2R^2}$$

CONVOLUTION

Frequently it is necessary to predict the outcome of a system before it is actually constructed. Of course, we can determine the transfer function of the system and then multiply the transfer function by the input signal. However, many times we do this by using the time response of the system and the time equation of the input signal. Such an operation when we "coil" together these time responses is known as "convolution." For example, suppose we desire to determine the convolution of a rectangular pulse as shown in Fig. 2-18 and the rectangular network response shown in Fig. 2-19. We will call the rectangular pulse $f_{(t)}$ and the network response $h_{(t)}$. The convolution of these two time functions is written $f_{(t)} * h_{(t)}$. The procedure of convolving the two functions is as follows:

1. Take one of the time responses, for example $f_{(t)}$, and fold it back along the x axis. Now starting with the $f_{(t)}$ touching the $h_{(t)}$ function at t = 0, start sliding the $f_{(t)}$ through $h_{(t)}$. As the right side of the folded $f_{(t)}$ moves to the right through time, the area under

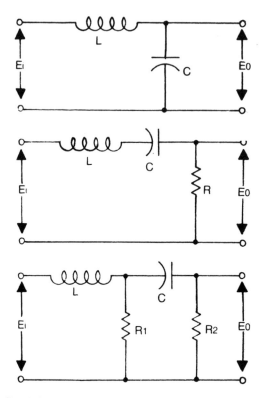

Fig. 2-17. Circuit for Problem 5.

the product of the two functions during the overlap is obtained. This is tabulated for the time overlap.

2. The procedure in step 1 is repeated and tabulations are made.

3. The area values are plotted versus time. For example, if the area for the overlap at 0.5 seconds is 100 then this value is plotted at time 0.5 seconds. All of the points are joined by a curve and an estimate of the resulting function occurs.

In Figs. 2-20 through 2-25 we see the process of convolution evolve. In Fig. 2-26 we see the convolved result. Note that the output is a trapezoidal response.

Example 2-15

A more interesting convolution example is that of a rectangu-

lar pulse of amplitude 1 volt and a network impulse response of
e^{-t}. Actually, a network whose impulse response is e^{-t} happens
to be a low pass filter as shown in Fig. 2-27. Since e^{-t} indicates
a response of e where RC equals one second, then the cutoff frequency of the low pass filter happens to be one radian per second.

The first thing we must do is fold back the input pulse about
the x axis. We now begin to slide the pulse through the e^{-t} response. We calculate the are for the overlap as we did previously.
Notice that the product of $f_{(t)}$ and $h_{(t)}$ under the overlap can be broken up into a group of rectangles and triangles. This will approximate the area. Remember that the area of the rectangle is the
product of the height and base, while the area of a triangle is one
half the product of the height and base.

As an example, for an overlap of 0.2 the area appears as in Fig.
2-28. It is made of approximately a rectangle and triangle. The area
under the rectangle is (.2 sec) (.8 volts) while the area under the
triangle is 1/2(0.2 sec) (0.2 volts). Thus the total area is the sum of

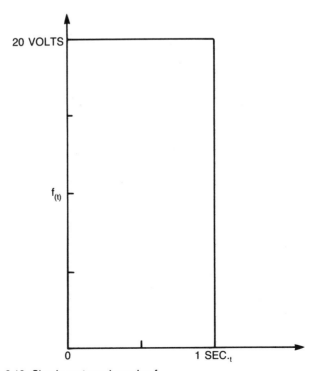

Fig. 2-18. Simple rectangular pulse f_t.

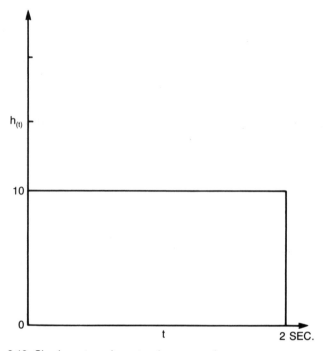

Fig. 2-19. Simple rectangular network response h_t.

the areas of .18. We repeat this as many times as we desire tabulating the area and the time. For example, in Fig. 2-29 we see the convolution at 1.0 sec. and in Fig. 2-30, the convolution at 2 sec.

Notice the result of this example (Fig. 2-31) is a waveform that starts rising from zero towards the maximum input size of 1 volt.

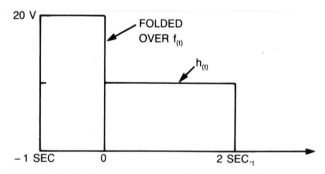

Fig. 2-20. The sliding-through process begins for convolution.

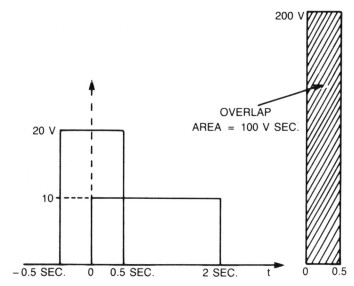

Fig. 2-21. Overlap at 0.5 seconds.

This appears to be an exponential rise. When the output pulse reaches a time *T* its value is maximum. At that time the input signal returns to zero. The output signal decays in an exponential man-

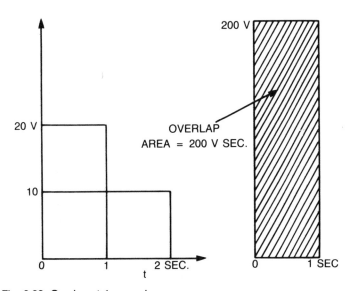

Fig. 2-22. Overlap at 1 second.

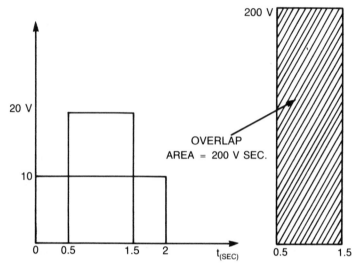

Fig. 2-23. Overlap at 1.5 second.

ner back towards zero. Actually, the response of this output signal can be written as in Equations 2-21 and 2-22:

Equation 2-21

$$v_{(t)} = V(1 - e^{-t}) \qquad 0 \le t \le r$$

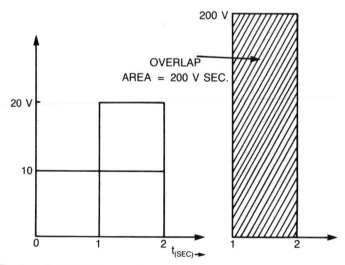

Fig. 2-24. Overlap at 2 seconds.

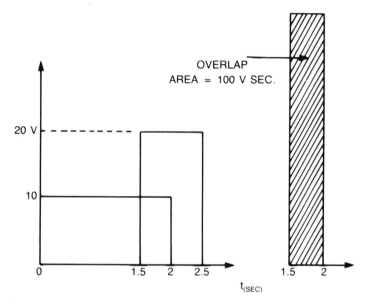

Fig. 2-25. Overlap at 2.5 seconds.

where V is the pulse amplitude.

At time r the value of $v_{(t)}$ becomes $V(1-e^{-r})$. For time equal or larger than r we can write Equation 2-22.

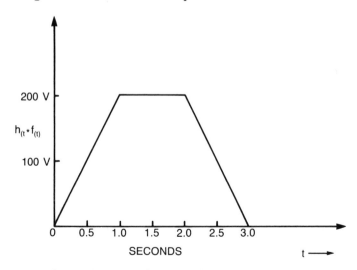

Fig. 2-26. Convolution result of a rectangular pulse and rectangular network response.

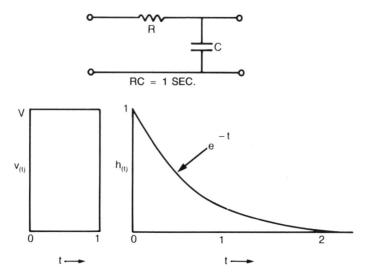

Fig. 2-27. A network whose impulse response is e^{-t} and a pulse of V volts and r width.

Equation 2-22

$$v_{(t)} = V(1 - e^{-r}) e^{-(t - r)} \quad r \leq t \leq \infty$$

These equations, of course, then represent the convoution performed.

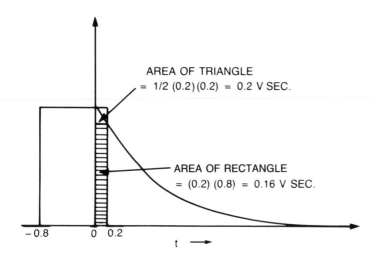

Fig. 2-28. Example 2-15 for an overlap area at 0.2 second.

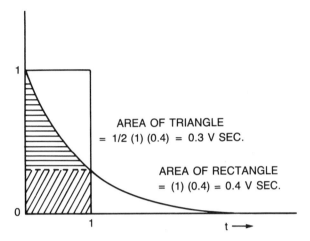

Fig. 2-29. Example 2-15 for an overlap area at 1 second.

We have just examined convolution by a graphical approach. Sometimes the more classical approach is used where equations are written for different time ranges. Convolution can also be carried out by a computer for the more difficult problems that occur. Convolution is a tool that is used frequently in signal and system analysis and the noise student should be exposed to it.

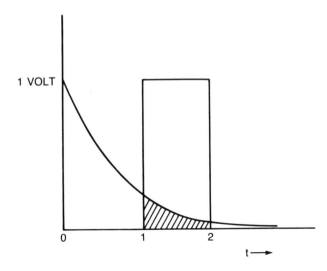

Fig. 2-30. Example 2-15 for an overlap area at 2 second.

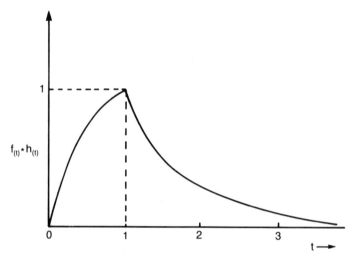

Fig. 2-31. Example 2-15 convolution result.

FOURIER TRANSFORM

Earlier in this chapter we discussed the Fourier series of a signal. The Fourier series describes a periodic signal. However, quite often we must deal with nonperiodic signals and their frequency response. One tool used in working with nonperiodic signals is the *Fourier Transform*. This is defined by Equation 2-23.

Equation 2-23

$$F_{(w)} = \int_{-\infty}^{\infty} f_{(t)} \, e^{-jwt} \, dt$$

$F_{(w)}$ is the Fourier Transform
$f_{(t)}$ is the time function of the signal in question
e^{-jwt} is the complex exponential function
w is radian frequency

To understand such a formula it is best to work several examples.

Example 2-16

Determine the Fourier Transform for the signal shown in Fig. 2-32. First of all, we define the time function:

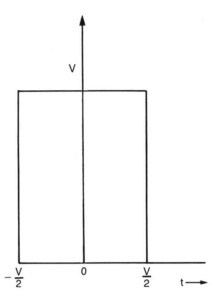

Fig. 2-32. Pulse for example 2-16.

$$f_{(t)} = V \qquad \frac{-r}{2} \leq t \leq \frac{r}{2}$$

$$f_{(t)} = 0 \text{ elsewhere}$$

Then we determine the Transform

$$F_{(w)} = \int_{-\infty}^{\infty} f_{(t)} e^{-jwt} dt = \int_{-\frac{r}{2}}^{\frac{r}{2}} V e^{-jwt} dt$$

$$= V \left[\frac{e^{-jwt}}{-jw} \right]_{-\frac{r}{2}}^{\frac{r}{2}}$$

$$= V \left[\frac{e^{-j \frac{wr}{2}} - e^{j \frac{wr}{2}}}{-jw} \right]$$

By dividing through by -1, we obtain,

$$F_{(w)} = \frac{V}{w} \frac{e^{j\frac{wr}{2}} - e^{-j\frac{wr}{2}}}{j}$$

We can very easily turn this equation into a sinusoidal term by multiplying numerator and denominator by two. This then becomes:

$$F_{(w)} = \frac{2V}{w} \frac{e^{j\frac{wr}{2}} - e^{-j\frac{wr}{2}}}{2j}$$

$$= \frac{2V}{w} \sin\frac{wr}{2}$$

We may turn this expression into a $\frac{\sin x}{x}$ term by multiplying numerator and denominator by $\frac{r}{2}$. The result becomes:

$$F_{(w)} = \frac{2V}{w} \frac{\left(\frac{r}{2}\right)}{\left(\frac{r}{2}\right)} \sin\frac{wr}{2}$$

$$F_{(w)} = V r \frac{\sin\frac{wr}{2}}{\frac{wr}{2}}$$

In Fig. 2-33 we see $F_{(w)}$ plotted versus frequency. We note that the first crossover point occurs when $\sin\frac{wr}{2}$ becomes zero. This of course equals zero at $w = 0$ but at that frequency we also obtain the denominator equal to zero and the limit of $\frac{\sin\frac{wr}{2}}{\frac{wr}{2}}$ as frequency approaches zero is unity. The next point that $\sin\frac{wr}{2}$ becomes zero

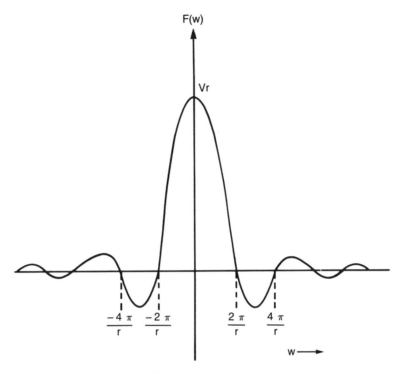

Fig. 2-33. Fourier transform illustration.

is when $\sin \dfrac{wr}{2} = \sin \pi$.

We can then solve for w:

$$\frac{wr}{2} = \pi$$

$$w = \frac{2\pi}{r}$$

Likewise, when $\dfrac{wr}{2} = -\pi$ the function goes to one also.

calculation of the transform is carried out by substitution of various values of $\dfrac{wr}{2}$ into the $F_{(w)}$ expression. For example, when $\dfrac{wr}{2}$

$= \dfrac{\pi}{2}$ we obtain

$$F_{(w)} = V r \; \frac{\sin \frac{\pi}{2}}{\frac{\pi}{2}}$$

$$= V r \; \frac{(1)}{\frac{\pi}{2}}$$

$$= \frac{2 V r}{\pi}$$

When $\dfrac{wr}{2} = \dfrac{-\pi}{2}$, we obtain

$$F_{(w)} = V r \; \frac{\sin\left[\frac{-\pi}{2}\right]}{\frac{-\pi}{2}}$$

$$= V r \; \frac{(-1)}{\frac{-\pi}{2}}$$

$$= \frac{2 V r}{\pi}$$

The values for $F_{(w)}$ are tabulated in Fig. 2-34.

w	F(w)
0	vr
$\pm \dfrac{\pi}{r}$	$\dfrac{2 \, vr}{\pi}$
$\pm \dfrac{2\pi}{r}$	0
$\pm \dfrac{3\pi}{r}$	$-\dfrac{2 \, vr}{2 \, \pi r}$
$\pm \dfrac{4\pi}{r}$	0

Fig. 2-34. Values for F_w in example 2-16.

Example 2-17

Determine the Fourier Transform for the Gaussian pulse shown in Fig. 2-35. First, we must define the time function.

$$f_{(t)} = V e^{\frac{-t^2}{2r^2}} \quad -\infty \leq t \leq \infty$$

Second, we determine the Fourier Transform:

$$F_{(w)} = \int_{-\infty}^{\infty} f_{(t)} e^{-jwt} dt = \int_{-\infty}^{\infty} V e^{-t^2/2r^2} e^{-jwt} dt$$

We may combine the exponential terms as follows:

$$= V \int_{-\infty}^{\infty} e^{\left[-\frac{t^2}{2r^2} + jwt \right]} dt$$

Now we complete the square on the exponent. It is known that:

$$a^2 + 2ab = (a + b)^2 - b^2$$

We can let $a = \dfrac{t}{\sqrt{2r}}$ and $2ab = jwt$

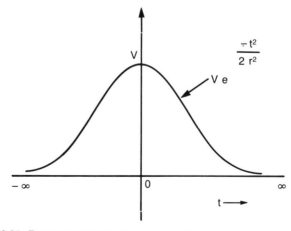

Fig. 2-35. Example 2-17 Gaussian pulse in time domain.

61

Therefore, $b = \dfrac{jwt}{2a} = \dfrac{jwt}{2\left(\dfrac{t}{\sqrt{2}\,r}\right)} = j\,\dfrac{wr}{\sqrt{2}}$

So, we can write:

$$\frac{t^2}{2r^2} + jwt = \left[\frac{t}{\sqrt{2}\,r} + \frac{wr}{j\sqrt{2}}\right]^2 - \frac{w^2r^2}{2}$$

Substitution of the completed square into the Fourier Transform:

$$F_{(w)} = V\int_{-\infty}^{\infty} e^{-\left(\left[\frac{t^2}{2r^2} + jwt - \frac{w^2r^2}{2}\right] + \frac{w^2r^2}{2}\right)} dt$$

This may be rewritten as:

$$F_{(w)} = V\int_{-\infty}^{\infty} e^{-\left(\frac{t^2}{2r^2} + jwt - \frac{w^2r^2}{2}\right)}\left(\frac{w^2r^2}{e-2}\right) dt$$

We can pull the term $e^{-\frac{w^2r^2}{2}}$ outside of the integral as a constant.

$$F_{(w)} = Ve^{-\frac{w^2r^2}{2}}\int_{-\infty}^{\infty} e^{-\left[\frac{t}{\sqrt{2}\,r} + \frac{jwr}{\sqrt{2}}\right]^2} dt$$

The integral is now in the form

$$F_{(w)} = Ve^{-\frac{w^2r^2}{2}}\int_{-\infty}^{\infty} e^{-x^2}\,dx$$

where

$$x = \frac{t}{\sqrt{2}\,r} + \frac{jwr}{\sqrt{2}}$$

and

$$\frac{dx}{dt} = \frac{d}{dt}\left[\frac{t}{\sqrt{2}\,r} + \frac{jwr}{\sqrt{2}}\right] = \frac{1}{\sqrt{2}\,r}$$

$$dx = dt\left(\frac{1}{\sqrt{2}\,r}\right)$$

Now we can write,

$$F_{(w)} = \frac{Ve^{-\frac{w^2 r^2}{2}}}{\frac{1}{\sqrt{2}\,r}} \int_{-\infty}^{\infty} e^{-x^2}\, dt\left(\frac{1}{\sqrt{2}\,r}\right)$$

$$F_{(w)} = V\sqrt{2}\,r\,e^{-\frac{w^2 r^2}{2}} \int_{-\infty}^{\infty} e^{-x^2}\, dx$$

$$F_{(w)} = V\sqrt{2}\,r\,e^{-\frac{w^2 r^2}{2}}\,(\sqrt{\pi})$$

$$F_{(w)} = \sqrt{2\pi}\,r\,V\,e^{-\frac{w^2 r^2}{2}}$$

We may now place various values of r into the Fourier Transform and calculate the various values for plotting the Transform. In Fig. 2-36 we see the Gaussian pulse's Fourier Transform. Notice that the frequency response is also Gaussian in shape. Note that at a value of $w = \frac{1}{r}$ the transform has a value of about 0.61 $\sqrt{2\pi}\,r\,V$.

CORRELATION FUNCTIONS

Over the years, research has resulted in the production of correlation methods that help in the removal of signals from noise. One type of correlation is called autocorrelation. In this correlation process, a periodic signal is compared with a shifted version of the signal itself. The product of the signal $x_{(t)}$ and the shifted version $x_{(t-r)}$ is evaluated at many different times and the results added together and averaged to form the correlation function. Such a function depends on the time r as well as the shape of the signal.

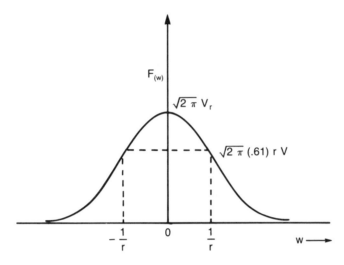

Fig. 2-36. Gaussian pulse in frequency domain.

The formula for the autocorrelation function [Auc(r)] is given by:

$$Auc\ (r) = limit\ \frac{1}{T} \int_0^T x_{(t)}\ x_{(t-r)}\ dt$$

$$T \rightarrow \infty$$

As r is varied the autocorrelation function takes on different values. When r equals the period, we may write:

$$Auc\ (r) = limit\ \frac{1}{T} \int_0^T x_{(t)}^2 dt$$

$$T \rightarrow \infty$$

This formula likewise holds when r equals an integer number of periods.

Example 2-18

In Fig. 2-37 we see a square wave with a period T. Suppose we correlate the signal with $r = T$. The formula becomes:

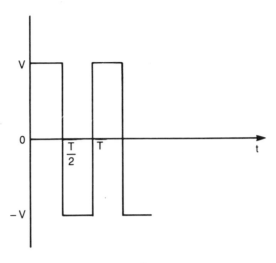

Fig. 2-37. Square wave with a period T.

$$Auc\ (r) = limit\ \frac{1}{T} \int_{0}^{\frac{T}{2}} v^2\ dt\ +\ \frac{1}{T} \int_{\frac{T}{2}}^{T} (-V)^2\ dt$$

$$= V^2$$

If we allow $r = \dfrac{T}{2}$ then, we obtain the following:

$$Auc\ (r) = limit\ \frac{1}{T} \int_{0}^{\frac{T}{2}} (V)(-V)\ dt\ +\ \frac{1}{T} \int_{\frac{T}{2}}^{T} (V)(-V)\ dt\ =\ +V^2$$

$$T \to \infty$$

Should we allow r to become $\dfrac{T}{4}$ we have a situation as below.

$$Auc = lim\ \frac{1}{T} \int_{0}^{\frac{T}{4}} (V)(-V)\ dt\ +\ \frac{1}{T} \int_{\frac{T}{4}}^{\frac{T}{2}} (V)(V)\ dt\ +\ \frac{1}{T} \int_{\frac{T}{2}}^{\frac{3T}{4}} (V)(-V)\ dt$$

$$T \to \infty$$

$$+ \frac{1}{T} \int_{\frac{3T}{4}}^{T} (-V)(V) \, dt$$

$$= \quad 0$$

If we continue this procedure and plot the resulting function we obtain the autocorrelation function as shown in Fig. 2-38. Note that it is triangular in nature.

Should we find the autocorrelation function of a white noise process we find it becomes an exponential decaying function. Obviously, when the noise signal is compared with itself at $r = 0$ there is only naturally maximum correlation. As r increases less correlation occurs. In Fig. 2-39 we see the autocorrelation function of a white noise process.

With crosscorrelation, a periodic signal is compared with a different signal for different values of r.

The crosscorrelation [Cc(r)] function is given by:

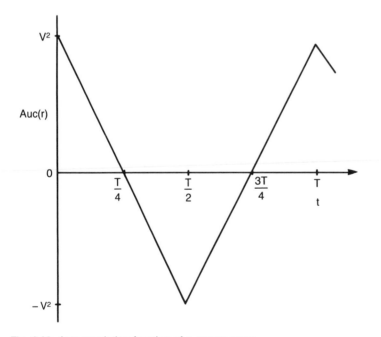

Fig. 2-38. Autocorrelation function of a square wave.

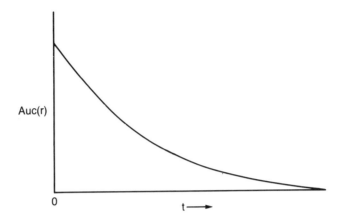

Fig. 2-39. Autocorrelation function of white noise.

$$Cc\ (r) = \lim_{T \to \infty} \frac{1}{T} \int_0^T X\ (t)\ Z\ (t-r)\ dt$$

where $Z(t-r)$ is the second signal.

A very special case of crosscorrelation is known as signal averaging. When a signal plus noise is repeatedly averaged the noise contributions get smaller and eventually equal to zero.

Anyone planning a career in noise measurement must become familiar with correlation techniques.

DECIBELS

Many times it is necessary to talk of power and voltage ratios that may have very large differences. When comparing these it may be more convenient to compare a linear measure of the ratio. This can be seen many times in nature. For example, let us say that we move from a room where there is nearly no sound to a very noisy environment where the sound is one million times the original "nearly quiet" condition. The ear does not respond linearly. It will respond in a logarithmic manner that allows us to hear the sound at about sixty times the original. The same thing occurs for our eyes. A light that is one thousand times the power of another light is only seen as thirty times as bright to us.

A power ratio may be written in a logarithmic form by the following formula given by Equation 2-26.

67

Equation 2-26

$$\#Bels = \log_{10} \frac{P_2}{P_1}$$

where P_2 and P_1 are respective power values. The unit is given in Bels. To convert these to smaller units we write Equation 2-27.

Equation 2-27

$$\#dB = 10 \log_{10} \frac{P_2}{P_1}$$

where the unit is measured in decibels. (1 decibel is one tenth a Bel)

Example 2-19

As an example, a power ratio of 20 would have a dB (decibel) value of

$$
\begin{aligned}
\#dB &= 10 \log 20 \\
&= 10 \ (1.3) \\
&= 13 \ dB
\end{aligned}
$$

Corresponding values of dB and power ratios are given in Fig. 2-40.

POWER RATIOS	#dB
10^{-5}	-50
10^{-4}	-40
10^{-3}	-30
10^{-2}	-20
10^{-1}	-10
$10^0 = 1$	0
10^1	10
10^2	20
10^3	30
10^4	40
10^5	50

Fig. 2-40. dB and power ratios.

Power Gain and Voltage Gain in dB

Consider the system shown in Fig. 2-41. The input voltage is V_i, the input resistance is R_i, the output voltage is V_o and the load resistance is R_L.

To determine the dB value of the power gain we write Equation 2-28.

Equation 2-28

$$\#dB = 10 \log \frac{P_o}{P_i}$$

$$= \log \frac{\dfrac{Vo^2}{R_L}}{\dfrac{Vi^2}{R_i}}$$

$$= 10 \log \frac{Vo^2}{Vi^2} \left(\frac{R_i}{R_L} \right)$$

This can be written as Equation 2-29.

Equation 2-29

$$\#dB = 20 \log \frac{V_o}{V_i} + 10 \log \frac{R_i}{R_L}$$

The first part of the power gain is the voltage gain while the second part of the power gain is the resistance correction factor.

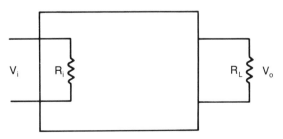

Fig. 2-41. System to explain dB.

Example 2-20

Determine the power gain for the system when $V_0 = 40$ volts,

$$V_i = 2 \text{ volts}, R_i = 1 \, k\,\Omega \text{ and } R_L = 100 \, \Omega.$$

$$\#dB = 20 \log \frac{40 \, V}{2 \, V} + 10 \log \frac{1 \, k}{0.1 \, k}$$

$$= 20 \log 20 + 10 \log 10$$

$$= 20 \,(1.3) + 10 \,(1)$$

$$= 36 \, dB$$

Example 2-21

Determine the power gain for the system when $V_0 = 40$ volts,

$$V_i = 2 \text{ volts}, R_i = 1 \, k\,\Omega \text{ and } R_L = 10 \, k\,\Omega$$

$$\#dB = 20 \log \frac{40 \, V}{2 \, V} + 10 \log \frac{1k}{10 \, k}$$

$$= 20 \log 20 + 10 \log \frac{1}{10}$$

$$= 20 \,(1.3) + 10 \,(-1)$$

$$= 16 \, dB$$

We note that if the input resistance and load resistance are equal, then the resistance correction factor becomes zero and the power gain in dB is the same value as the voltage gain in dB.

Example 2-22

Determine the power gain for the system when $V_0 = 40$ volts,

$$V_i = 2 \text{ volts}, R_i = 1 \, k\,\Omega \text{ and } R_L = 1 \, k\,\Omega.$$

$$\#dB = 20 \log \frac{40}{2} + 10 \log \frac{1 \, k}{1 \, k}$$

$$= 20 \log 20 + 10 \log 1$$

$$= 20 \ (1.3) + 10 \ (0)$$

$$= 26 \ dB$$

dBm

The dBm is a unit to describe a power ratio when the reference is one milliwatt. To convert a power of P watts to dBm we use Equation 2-30.

Equation 2-30

$$\#dBm = 10 \log \frac{P}{1 \times 10^{-3}}$$

where P is in watts

or equation 2-31,

Equation 2-31

$$\#dBm = 10 \log \frac{P}{1}$$

where P is in milliwatts

Example 2-23

Convert 4 watts to dBm.

$$\#dBm = 10 \log \frac{4}{1 \times 10^{-3}}$$

$$= 36.02 \ dBm$$

dBW

The dBW is a unit to describe a power ratio when the reference is one watt. To convert a power of P watts to dBW we use Equation 2-32.

Equation 2-32

$$\#dBW = 10 \log \frac{P}{1}$$

where P is in watts.

Example 2-24

Convert 16 mW to dBW.

$$\#dBW = 10 \log \frac{16 \times 10^{-3}}{1}$$

$$= 10 \log \quad 16 \times 10^{-3}$$

$$= -17.95 \; dBW.$$

dBV

The dBV is a unit used to describe a power ratio when the reference is one volt. To convert a voltage of V volts to dBV we use Equation 2-33.

Equation 2-33

$$\#dBV = 20 \log \frac{V}{1}$$

Example 2-25

Convert 32 volts to dBV.

$$\#dBV = 20 \log \frac{32}{1}$$

$$= 20 \log 32$$

$$= 30.1 \; dBV.$$

SUMMARY

1. When energy propagates through a network it is represented by a signal.

2. Signals may be periodic, non periodic and random.

3. The important signal parameters are the average level, rms level, form factor, and crest factor.

4. A periodic signal can be represented by a Fourier series. This is an average level and a group of sine and cosine waveforms.

5. Although we have studied the frequency makeup of a periodic signal, we realize the similarity to the frequency components in a noise signal.

6. In all electronic problems we encounter filtering and operations. When dealing with noise we are especially interested in the low pass characteristics of the filter.

7. The transfer function of a filter describes how the ratio of output to input ratio varies with frequency.

PROBLEMS AND QUESTIONS

1. Discuss the five types of filters mentioned in this chapter.

2. For the waveform in Fig. 2-2A, compute the average and rms level. Use a pulse height of 40 volts.

3. For the waveform in Fig. 2-2A, compute the form factor and crest factor.

4. Evaluate the Fourier series for the waveform in Fig. 2-3 and draw the frequency spectrum.

5. For each of the circuits shown in Fig. 2-17, evaluate the transfer function.

6. For equations 2-17, 18, 19, and 20 in the text determine the magnitude of the transfer function.

7. What is the duty cycle for the waveform shown in Fig. 2-42? Compute the duty cycle.

8. Determine the average rms level for the waveform shown in Fig. 2-43.

9. A square wave has a Fourier series given by the following equation:

$$e = \frac{4V}{\pi} \left[\cos 1000t - \frac{1}{3} \cos 3000t + \frac{1}{5} \cos 5000t - \frac{1}{7} \cos 7000 \ t \right]$$

a) What is the square wave period?

b) Draw the frequency spectrum.

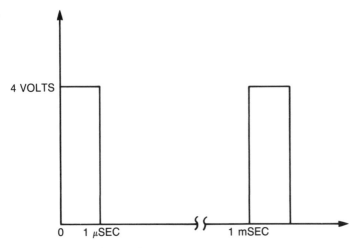

Fig. 2-42. Waveform for problem 7.

c) If *V* is 10 volts, draw the time diagram for the square wave.

10. A sawtooth wave has a Fourier series given by the following equation:

$$e = \frac{2A}{\pi} \left[\cos 1000t - \frac{1}{2} \cos 2000t + \frac{1}{3} \cos 3000\ t - \frac{1}{4} \cos 4000\ t \right]$$

Fig. 2-43. Waveform for problem 8.

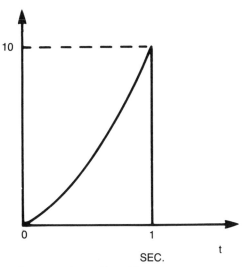

Fig. 2-44. Response for problem 11.

Draw the frequency spectrum for the sawtooth wave.

11. Convolve the time responses shown in Figs. 2-44 and 2-45. Use graphical methods.

12. Find the Fourier Transform for the pulses shown in Figs. 2-46 and 2-47.

13. Sketch the autocorrelation function for the waveform shown in Fig. 2-48.

14. Evaluate the phase and magnitude of the following transfer function at w = 5 rad/sec.

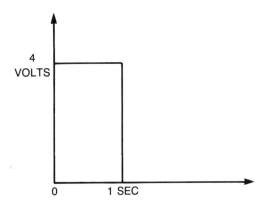

Fig. 2-45. Waveform for problem 11.

$$|H(jw)| = \frac{10}{\sqrt{(w)^4 + 20(w)^2 + 64}}$$

15. Evaluate the Fourier Series for the waveform shown in Fig. 2-49.

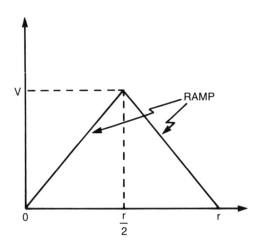

Fig. 2-46. Waveform for problem 12.

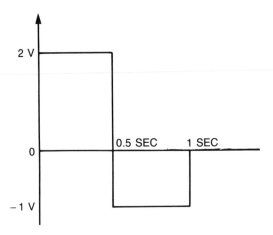

Fig. 2-47. Waveform for problem 12.

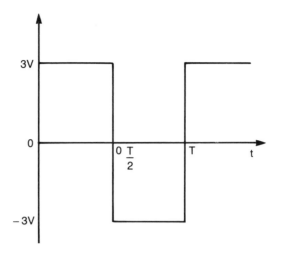

Fig. 2-48. Waveform for problem 13.

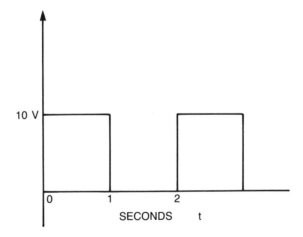

Fig. 2-49. Waveform for problem 15.

Chapter 3

Thermal Noise

I N CHAPTER ONE WE BRIEFLY INVESTIGATED THE SUBJECT OF thermal and shot noise. In this chapter we will look at the subject of thermal noise in greater detail and we will learn how to make calculations which predict the size of these noise components in circuits composed of resistors.

THERMAL NOISE IN RESISTORS

Thermal noise is caused by the random motion of free electrons in a resistive material. This type of noise is also called resistance noise or Johnson noise. In 1928 J. B. Johnson showed that the noise power of a resistor, which was caused by thermal agitation, had a linear variation with measuring equipment bandwidth.

As the temperature of a resistor or resistive material increases, the agitation of the electrons becomes more vigorous, raising the noise power and its associated noise voltage. Should the temperature decrease, then the noise will decrease. If the temperature decreases to absolute zero, which is – 273 degrees Centigrade, the thermal noise will fall to zero.

In Fig. 3-1 we see a resistor and its associated thermal noise. Note that electrons move in directions that can be broken down into two components—one which is vertical in motion and parallel

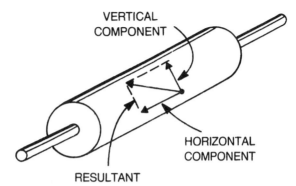

Fig. 3-1. Horizontal and vertical components for a thermally agitated electron in a resistor.

to the resistor cross section and one which is horizontal in motion and perpendicular to the resistor cross section. Since the resultant of these components flows through small portions of the resistance, a small voltage drop occurs in these small portions. At any instant these individual voltage drops combine, giving a total noise voltage across the resistor terminals. The various drops differ in polarity and value from instant to instant and give a varying noise across the terminals.

Over a long period of time it can be seen that the area above the zero reference is the same as the area under the zero line. From Chapter 2 we learned that the average voltage of a waveform depended on the area under the waveform. If the area under the waveform totals zero, there is *no* average level. Thus for thermal noise generated in a resistor, there is no average voltage level except zero. Refer to Fig. 3-2.

Factors Determining Thermal Noise

The noise produced by a resistor due to the thermal effect is due to resistor value in ohms, temperature in degrees Kelvin and bandwidth in Hertz.

Thermal noise is not dependent on frequency (except at extremely high frequencies—around 10^{13} Hertz and above) but is dependent on bandwidth. It has been determined that the available noise power due to the thermal effect in a resistor is:

Equation 3-1

$$P_a = kTB$$

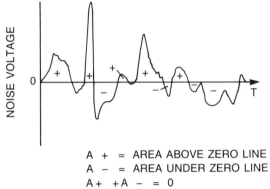

A + = AREA ABOVE ZERO LINE
A − = AREA UNDER ZERO LINE
A+ +A − = 0

Fig. 3-2. Thermal noise voltage in a resistor. The sum of the area above zero line and below zero line equals zero.

where k = Boltzmanns constant,

$$(1.38 \times 10^{-23} \text{ Joules/degree Kelvin})$$

T = Temperature in degrees Kelvin,

B = Bandwidth in question in Hertz.

A resistor can be thought of as being noisy or being noiseless with a noise generator in series with it. This is shown in Fig. 3-3.

Thermal Noise Voltage

Suppose we want to know how much power can be delivered to an equal value resistor connected to our noisy resistor. We use the circuit shown in Fig. 3-4. The power available from any gener-

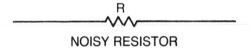

NOISY RESISTOR

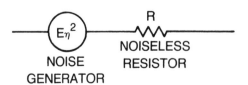

NOISE GENERATOR   NOISELESS RESISTOR

Fig. 3-3. A noisy resistor is equivalent to a noiseless resistor in series with a noise generator.

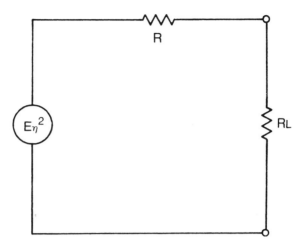

Fig. 3-4. Circuit to calculate available noise power from resistor R.

ator is the square of the open circuit voltage divided by four times the generator's internal resistance.

Likewise, we could say that the available noise power from a resistor R is given by Equation 3-2.

Equation 3-2

$$P_a = \frac{E_n^2}{4R}$$

The symbol E_n is the rms thermal noise voltage.

Because $P_a = kTB$ we may make a substitution into Equation 3-2.

Equation 3-3

$$kTB = \frac{E_n^2}{4R}$$

Solving for E_n^2, we obtain

$$E_n^2 = 4kTRB$$

Equation 3-3 states that the square of the rms noise voltage of a resistor with value R is proportional to the product of its Kelvin temperature, resistance and bandwidth. The constant 4k makes the equation complete.

81

Thus, the rms noise voltage of a resistor R is given by Equation 3-4.

Equation 3-4

$$E_n = \sqrt{4kTRB}$$

Equation 3-4 states that to find the rms noise voltage across a resistor we take the square root of the product of four times Boltzmanns constant, resistance, temperature and bandwidth in question. It must be remembered that we must use the Kelvin temperature when calculating noise voltages. To convert Centigrade temperatures to Kelvin use Equation 3-5.

Equation 3-5

$$T = C + 273$$
where T = temperature in degrees Kelvin,
C = temperature in degrees Centigrade.

To convert Fahrenheit temperatures to Kelvin use Equation 3-6.

Equation 3-6

$$T = 5/9\ (F - 32) + 273$$
where F = temperature in degrees Fahrenheit.

Now let us work a problem with thermal noise.

Example 3-1

Compute the thermal noise voltage across a resistor if the resistor is at a temperature of 17 degrees Centigrade, the resistor value is one Megohm and the bandwidth of interest is 500 kHz.

$$E_n^{\ 2} = 4 \times k \times T \times R \times B$$

$$E_n^{\ 2} = 4 \times 1.38 \times 10^{-23} \times (17 + 273) \times 1 \times 10^6 \times 5 \times 10^5$$

$$E_n^{\ 2} = 80 \times 10^{-10}$$

$$E_n = 8.95 \times 10^{-5} = 89.5\mu V.$$

Equivalent Circuits For Resistors

A resistor may be modeled in a voltage generator equivalent

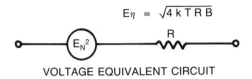

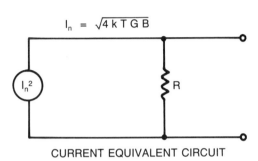

VOLTAGE EQUIVALENT CIRCUIT

CURRENT EQUIVALENT CIRCUIT

Fig. 3-5. Voltage and current equivalent circuits for a resistor.

circuit or on a current generator equivalent circuit. These circuits are shown in Fig. 3-5. Since it is somewhat easier to work with the square of the voltage or current, the generator has been marked I_n^2 or E_n^2.

For the current equivalent circuit, the square of the rms current is calculated by shorting the end terminals of the voltage equivalent, giving a value of I_n^2 equal to E_n^2/R^2. We can then write Equation 3-7.

Equation 3-7

$$I_n^2 = \frac{E_n^2}{R^2} = \frac{4kTRB}{R^2} = \frac{4kTB}{R}$$

We can thus write,

Equation 3-8

$$I_n^2 = 4kTGB$$

where G equals the conductance or 1/R.

THERMAL NOISE RELATED TO ENERGY

Any resistor has a shunt capacity associated with it. Across the terminals of the resistor we have a noise voltage. The energy associated with the voltage variation is stored in the electric field of

83

the capacitance. The energy is given by Equation 3-9:

Equation 3-9

$$W = 1/2CE_n^2$$

where W is the energy in joules,

C is the capacitance in farads,

E_n is the rms noise voltage in volts.

From physics it is known that if the resistor and capacitor are at thermal equilibrium at a temperature of T, the average energy stored in the capacitor must be $1/2\,kT$ which is one half times the product of Boltzmanns constant and the temperature in degrees Kelvin. Therefore we may write Equation 3-10.

Equation 3-10

$$1/2\,kT = 1/2CE_n^2$$

Solving for E_n^2 we have,

Equation 3-11

$$E_n^2 = kT/C$$

Thus the square of the rms noise voltage across a resistor is proportional to Boltzmanns constant and Kelvin temperature and inversely proportional to capacitance.

We know that the parallel RC circuit shown in Fig. 3-6 will have a certain frequency when the absolute magnitude of its capacitive reactance equals the resistance. That is,

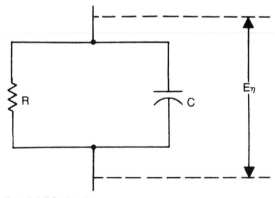

Fig. 3-6. Parallel RC circuit.

$$\frac{1}{2\pi f C} = R.$$

Solving for f we find

Equation 3-12

$$f_c = \frac{1}{2\pi RC}$$

This frequency represents the 3 dB bandwidth of the circuit and is called the "cutoff frequency". I will call this frequency "f_c". If we solve the equation for capacitance in terms of bandwidth we obtain

$$C = \frac{1}{2R(f_c)\pi}$$

Letting $f_c = B$, $C = \dfrac{1}{2RB\pi}$

Substituting this formula into Equation 3-11 we obtain

Equation 3-13

$$E_n^2 = \frac{kT}{C} = \frac{kT}{\dfrac{1}{2RB\pi}} = 2kTRB\pi$$

We notice that the noise given by Equation 3-3 is smaller than that given by Equation 3-13. The reason is that the true noise bandwidth is not $1/2\pi RC$. Instead, it is 1/4 RC. This will be shown at a later time in this book. If we solve B = 1/4 RC for C, we get C = 1/4 RB. The noise then becomes,

$$E_n^2 = \frac{kT}{C} = \frac{kT}{\dfrac{1}{4RB}} = 4kTRB.$$

This is the same formula as was given for Equation 3-3.

BASIC STATISTICAL INFORMATION

Johnson or resistance noise varies according to a Gaussian distribution which follows a probability function as given by Equation 3-14.

Equation 3-14

$$P\,(e_n) = \frac{1}{\sqrt{2\pi E_n^2}}\ \epsilon^{\dfrac{-e_n^2}{2\,E_n^2}}$$

where $P\,(e_n)$ is the spot probability of the noise voltage having a value e_n,

$E_n^2 \quad = \quad$ variance or rms noise voltage squared,

$E_n^2 \quad = \quad 4kTRB.$

If we plot $P\,(e_n)$ versus e_n we obtain the probability density diagram. This is shown in Fig. 3-7. It is easily seen from the diagram that small voltage variations are more likely to occur than large variations. Also, remember that noise voltage variations are both

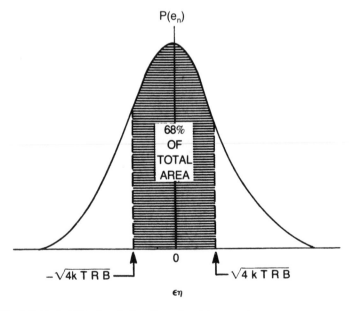

Fig. 3-7. Probability density function of resistance noise.

positive and negative. The probability that the noise voltage varies between $-\sqrt{4kTRB}$ and $\sqrt{4kTRB}$ is about 0.68, and the probability that the noise voltage will be larger than those values is about 0.32. Also, it is unlikely that any noise voltage will be larger than $-3\sqrt{4kTRB}$ or $3\sqrt{4kTRB}$. It is almost impossible that any voltage will exceed $-4\sqrt{4kTRB}$ or $4\sqrt{4kTRB}$.

NOISE COMPUTATIONS

Because of the nature of this noise, it is not possible to add several noise voltages or currents together. Instead you must add the squares of the individual noise voltages or currents together, and then extract the square root to find the overall noise voltage or current. This is shown in Equation 3-15.

Equation 3-15

$$E_{nT} = \sqrt{E_{n1}^2 + E_{n2}^2 + \cdots - E_{nn}^2}$$

Example 3-2

Three resistors, each producing rms thermal noise voltages of 20 μV, 30 μV, and 40 μV respectively, are placed in series. What is the total rms thermal noise voltage across the terminals of the series circuit?

$$E_{nT} = \sqrt{(20 \ \mu V)^2 + (30 \ \mu V)^2 + (40 \ \mu V)^2}$$

$$E_{nT} = \sqrt{(400 + 900 + 1600) (\mu V)^2}$$

$$E_{nT} = \sqrt{2900 (\mu V)^2}$$

$$E_{nT} = 53.8 \ \mu V.$$

Suppose we were not given the individual noise voltages but were given resistance values, bandwidth and temperature. Then the formula would be written as follows for two resistors in series.

Equation 3-16

$$E_n = \sqrt{E_{n1}^2 + E_{n2}^2} = \sqrt{4kTR_1B + 4 \ TR_2B}$$

Equation 3-16 can be rewritten as

$$E_n = \sqrt{4 \ k \ T \ B \ (R_1 + R_2)}$$

where $R_1 + R_2$ represents the total series resistance. In general, for resistors in series, the total rms noise voltage is

Equation 3-17

$$E_n = \sqrt{4kTBR_T}$$

where R_T is the total resistance of the resistors in series.

Example 3-3

Refer to Fig. 3-8. Calculate the total rms noise voltage in a bandwidth of 1 MHz. The temperature is 300 degrees K.

1. Calculate the total resistance:

$$R_T = R_1 + R_2 + R_3$$
$$= 30\ k + 40\ k + 50\ k = 120\ k\Omega$$

2. Apply formula $E_n = \sqrt{4kTBR_T}$

$$E_n^2 = 4 \times 1.38 \times 10^{-23} \times 3 \times 10^2 \times 10^6 \times 1.2 \times 10^5$$

$$E_n = \sqrt{19.8 \times 10^{-10}} = 4.45 \times 10^{-5}$$
$$E_n = 44.5\ \mu V.$$

Example 3-4

Refer to Fig. 3-9. Calculate the total rms noise voltage in a bandwidth of 2 MHz. The temperature is 17 degrees Centigrade.

1. Calculate the total resistance

$$R_T = \frac{R_1\ (R_2 + R_3)}{R_1 + R_2 + R_3}$$

Fig. 3-8. Circuit for Example 3-3.

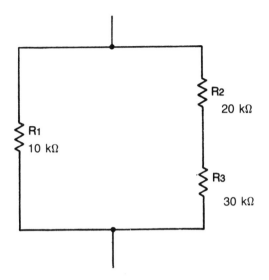

Fig. 3-9. Circuit for Example 3-4.

$$= \frac{10\ k\ (20\ k\ +\ 30\ k)}{10\ k\ +\ 20\ k\ +\ 30\ k}$$

$$= 8.33\ k\Omega$$

2. Calculate the temperature in degrees Kelvin.

$$T\ =\ 273\ +\ C$$
$$T\ =\ 273\ +\ 17\ =\ 290\ \text{degrees}\ K.$$

3. Find total rms noise voltage.

$$E_n{}^2\ =\ 4\ k\ T\ R_T\ B$$
$$=\ 4\ \times\ 1.38\ \times\ 10^{-23}\ \times\ 2.9\ \times\ 10^2\ \times\ 8.33\ \times\ 10^3\ \times\ 2\ \times 10^6$$
$$E_n\ =\ 16.3\ \mu V.$$

The spectral density of thermal noise is considered "white" in nature. To calculate the density we take Equation 3-3 and re-arrange it so we will have rms noise voltage squared per bandwidth. This gives rms noise voltage squared per Hz as given by Equation 3-18. Namely, the spectral density is

Equation 3-18

$$G_{v(f)} = \frac{E_n^{\,2}}{B} = \frac{4\,kTRB}{B} = 4kTR$$

Example 3-5

Find the spectral density of a thermal noise process with rms noise voltage squared equal to 15×10^{-6} volts^2 and bandwidth of 5×10^4 Hz.

$$G_{v(f)} = \frac{E_n^{\,2}}{B} = \frac{15 \times 10^{-6} \text{ Volts}^2}{5 \times 10^4 \text{ Hz}} = 3 \times 10^{-10} \frac{\text{Volts}^2}{\text{Hz}}$$

Example 3-6

Find the spectral density of a resistor of 2 M Ω, temperature of 300 degrees K, and bandwidth of interest equal to 1 MHz.

$$G_{v(F)} = 4\,k\,T\,R = 4 \times 1.38 \times 10^{-23} \times 3 \times 10^2 \times 2 \times 10^6$$
$$= 3.31 \times 10^{-14} \text{ V}^2/\text{Hz}.$$

THERMAL NOISE AT HIGH FREQUENCIES

The formulas we have been using for thermal noise cannot be valid for extremely large frequencies. Although this can only be explained with quantum mechanics, we can attempt to look at this from the condition of infinite frequency. Placing infinity in the equation for noise squared voltage for the bandwidth gives $E_n^{\,2} = 4\,kTR(\infty)$. This implies that noise squared voltage is infinite. But it is not possible since it has been shown that for very high frequencies, the spectral density begins to become a function of frequency. The density actually falls off with frequency as shown in Fig. 3-10.

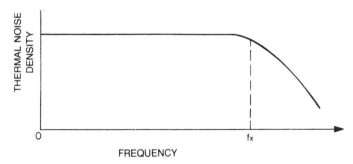

Fig. 3-10. Spectral density begins to roll off at high frequencies.

A noticeable decrease in the spectral density occurs for a temperature of 290 degrees Kelvin at about 6×10^{12} Hz.

To obtain an estimate for the highest frequency a thermal noise process can still be considered flat, use Equation 3-19 where fx is that frequency.

Equation 3-19

$$f_x = .15 \; k \; T \times 10^{34} \text{ Hz}$$

Example 3-7

A thermal noise process is being generated from a source of 100 degrees K. Compute the highest frequency for which the spectrum would be flat.

$$f_x = .15 \; kT \times 10^{34} = .15 \; (1.38 \times 10^{-23}) \; (1 \times 10^2) \times 10^{34}$$
$$= .207 \times 10^{13} = 2.07 \times 10^{12} \text{ Hz.}$$

Thus at 2070 Gegahertz, the noise spectrum begins to roll off.

THE NOISY CUBE

A good exercise in calculating noise is to determine the noise voltage across the diagonal points a and b as defined by the cube as shown in Fig. 3-11.

There are many ways that one may try to solve this problem, however, the most straightforward method is to use the symmetrical properties of a cube which has equal resistors in all legs.

First of all, assume that a current I enters point a. Since there are three equal paths for current to flow from point a, each one of these currents can be written as $\dfrac{I}{3}$. At point b, these three currents again meet and leave the cube. Therefore, legs A, B, C, J, K, and L all carry currents $\dfrac{I}{3}$.

Due to the fact that there are six legs which are dissipating this power, the sum of these powers becomes $6 \left(\dfrac{I}{6}\right)^2 R$ or $\dfrac{I^2}{6} R$.

If we now sum the powers due to each leg we have:

$$\frac{(I)^2}{6} R + \frac{2}{3} (I)^2 R = P_T.$$

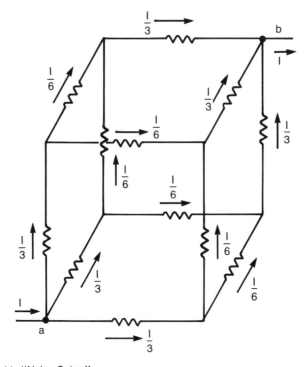

Fig. 3-11. "Noisy Cube."

The total power must be the product of the total resistance and the total current squared. Therefore we write

$$\frac{(I)^2}{6} R + \frac{2}{3} (I)^2 R = R_T (I)^2$$

Cancelling the $(I)^2$ terms and combining fractions, we obtain

$$\frac{3 R + 12 R}{6 \, (18)} = R_T$$

$$\frac{15}{18} R = R_T$$

$$R_T = \frac{5}{6} R$$

The noise voltage would then be given by the thermal noise formula

$$E_n = \sqrt{4\,k\,T\,R_T\,B}$$

If R was 1000 ohms, T 290 °K and B = 1 MHz, the noise would be found by the following procedure:

1. $R_T = \dfrac{5}{6} R = \dfrac{5}{6} (1000) = 833 \ \Omega$

2. $E_n = \sqrt{4\,(1.38 \times 10^{-23})\,(2.9 \times 10^2)\,(8.33 \times 10^2)\,(10^6)}$

   $= \sqrt{1.333 \times 10^{-11}}$

   $= \sqrt{13.33 \times 10^{-12}}$

   $= 3.65 \ \mu\text{V}$

SUMMARY

1. Thermal noise power is proportional to bandwidth, resistance and temperature.

2. Thermal noise voltage is proportional to the square root of bandwidth, resistance and temperature.

3. In thermal noise calculations we must use the temperature in degrees Kelvin.

4. The equivalent circuit for a noisy resistor can be presented in voltage or current form.

5. It is usually easier to work with the square of rms noise voltage or the square of rms noise current.

6. When computing the total noise voltage across a series circuit composed of resistors, we must take the square root of the sum of the squares of the individual noise voltages.

7. The spectral density for resistor noise due to the thermal effect is essentially constant. However, for very high frequencies, the spectral density begins to fall.

PROBLEMS AND QUESTIONS

1. Compute the thermal rms noise voltage in a 300 Ω resistor if the temperature is 17 degrees C and the bandwidth is 6 MHz.

2. Repeat problem 1 if the temperature is 51 degrees C.

3. A resistor producing thermal noise in a bandwidth of 7 kHz gives a noise voltage of 0.3 μV. What is the resistor value and the spectral density? Assume 17 degrees C.

4. Calculate the available power for a 1 M Ω resistor if the bandwidth is 10 kHz. The temperature is 300 °K.

5. Two resistors are producing rms noise voltage of 50 μV and 70 μV respectively. What is the total rms noise voltage if

    a. resistors are in series?

    b. resistors are in parallel?

6. If the shunt capacity of a resistor is 4pf, what is the total rms noise voltage? T = 300 °K

7. You are given a noisy resistor producing 60 μV of rms thermal noise. Draw the equivalent noise voltage and noise current circuits.

8. Calculate the total rms noise voltage across the circuit of Fig. 3-9. The temperature is 300 degrees K and the bandwidth of interest is 1 MHz.

9. A thermal noise process is being generated from a source of 50 degrees F. Compute the highest frequency for which the spectrum would be flat.

10. Make a plot of rms noise voltage versus bandwidth for a 10 K Ω resistor for bandwidths of 100 Hz, 200 Hz, 300 Hz, 400 Hz and 500 Hz. What can be said about the shape of the curve? T = 300 °K.

11. Discuss the results of varying temperatures for a resistor in regards to the thermal noise generated.

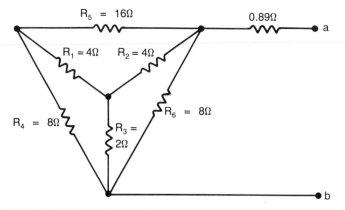

Fig. 3-12. Circuit for problem 3-13.

12. Prove that the area under the Gaussian curve is unity if the limits are from minus infinity and positive infinity.

13. Evaluate the noise across points a and b of the circuit of Fig. 3-12.

Chapter 4

Noise in
Electronic Devices

S O FAR WE HAVE LEARNED THAT RESISTORS PRODUCE THER-
mal noise. But what about devices like diodes and transistors?
We will cover these subjects in this chapter. We will learn about
the various types of noise in electronic devices and how to forecast the size of these noise signals.

SHOT NOISE

The term "shot noise" arose from the study of random variations in the emission of electrons from the cathode of a vacuum
tube. If these variations are amplified and listened to with a pair
of headphones or a loudspeaker, they sound like "lead shot" hitting a concrete wall. Shot noise has a flat spectral density like thermal noise. Therefore, shot noise can be considered a "white noise
process." Figures 4-1 and 4-2 show the appearance of shot noise
as a function of time and the spectral density respectively.

What Causes Shot Noise?

As mentioned in chapter one, when a direct current flows
through a device, there seems to be some random variations superimposed on this direct current. This variation is due to the random fluctuation in the emission of electrons from the emitting
surface. Also, velocity of electrons and the transit time of electrons

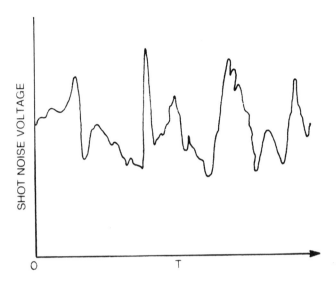

Fig. 4-1. Shot noise voltage versus time.

will play a role in determining the amount of shot noise. It is obvious that one major difference between shot and thermal noise is that shot noise is associated and dependent on a direct current. Thermal noise is not dependent on a direct current. Figure 4-3 compares shot and thermal noise.

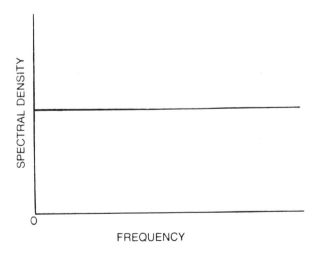

Fig. 4-2. Shot noise spectral density diagram.

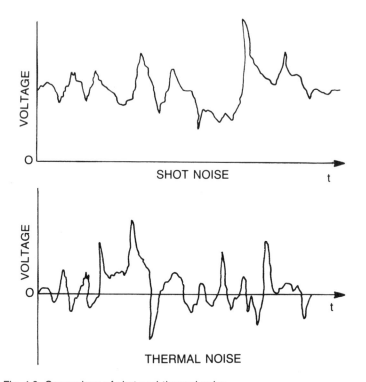

Fig. 4-3. Comparison of shot and thermal noise.

Factors Influencing Shot Noise

Shot noise is dependent on the following factors in both temperature limited vacuum tube diodes (a vacuum tube diode in which the emission from the cathode is limited by the cathode's temperature) and semiconductor p n junctions—direct component of current flow and bandwidth imposed on the device.

The square of the rms shot noise current is given by the following formula:

Equation 4-1

$$I_n^2 = 2eI_{dc}B$$

I_n is the shot noise component of current in amperes. I_{dc} is the direct current component flowing through the device in amperes.

e = charge on an electron or 1.6×10^{-19} coulombs.

B = bandwidth imposed on the electronic device in hertz.

Example 4-1

Compute the square of the rms noise current flowing through a temperature limited diode if I_{dc} is 10 ma, and B is 1 MHz.

$$I_n^2 = 2\,e\,I_{dc}B$$
$$= 2 \times 1.6 \times 10^{-19} \times 10 \times 10^{-3} \times 1 \times 10^6$$
$$= 3.2 \times 10^{-15} \text{ amperes}^2.$$

If we wanted to determine the rms shot noise current, we would extract the square root of I_n^2 as in Equation 4-2.

Equation 4-2

$$I_n = \sqrt{2eI_{dc}B}$$

Example 4-2

Using the value of I_n^2 from example one, calculate the rms shot noise current.

$$I_n = \sqrt{I_n^2}$$
$$= \sqrt{32 \times 10^{-16}}$$
$$I_n = 5.66 \times 10^{-8} \text{ amperes}$$
$$= 56.6 \text{ nano-amperes}$$

When shot noise current flows through a resistor of value R, a shot noise voltage develops across the resistor given by Equation 4-3.

Equation 4-3

$$E_{n\,shot} = (I_{n\,shot})\,(R)$$

Likewise, the square of the rms shot noise voltage across a

resistor is determined by multiplying the square of the rms shot noise current by the square of the resistance the shot current is flowing through. This is given by Equation 4-4.

Equation 4-4

$$E_{n\ shot}^2 = (I_{n\ shot}^2)\ (R^2)$$

Example 4-3

Compute the shot noise voltage across a 10 kΩ resistor if a shot noise current of 56.5 na flows through the resistor.

$$E_{n\ shot} = (I_{n\ shot})\ (R)$$
$$= (5.66 \times 10^{-8})\ (10^4)$$
$$= 5.66 \times 10^{-4}\ \text{volts}$$
$$= 566\ \text{microvolts}$$

Example 4-4

Compute the square of the rms noise voltage for Example 4-3.

$$E_{n\ shot}^2 = (I_{n\ shot}^2)\ (R^2)$$
$$E_{n\ shot}^2 = (5.66 \times 10^{-4})^2$$
$$= 32 \times 10^{-8}\ \text{volts}^2$$

Spectral Density of Shot Noise

To compute the spectral density of a shot noise process, we simply divide the square of the rms shot noise current or the square of the rms shot noise voltage by the bandwidth. Equations 4-5 and 4-6 indicate this.

Equation 4-5

$$G_{I\ shot} = \frac{i_{n\ shot}^2}{B} = \frac{2\ eI_{dc}B}{B}$$
$$= 2\ eI_{dc}$$

Equation 4-6

$$G_{V\,shot} = \frac{E_{n\,shot}^2}{B} = \frac{2eI_{dc}\,BR^2}{B}$$

$$= 2eI_{dc}R^2$$

We note that the units of the spectral densities are amps squared per hertz and volts squared per hertz respectively.

Example 4-5

A direct current flows through a resistor R. The direct current is 1 ampere and the current is flowing through a temperature limited diode which has a limiting bandwidth of 10 kHz. The resistor is 1 kΩ. Compute the spectral densities:

$$G_{I\,shot} = 2\,e\,I_{dc}$$

$$= 2 \times 1.6 \times 10^{-19} \times 1$$

$$= 32 \times 10^{-20} \text{ amperes}^2 \text{ per hertz}$$

$$G_{V\,shot} = 2\,e\,I_{dc}\,R^2$$

$$= 2 \times 1.6 \times 10^{-19} \times 1 \times (10^3)^2$$

$$= 3.2 \times 10^{-19} \times 10^6$$

$$= 3.2 \times 10^{-13}$$

$$= 32 \times 10^{-14} \text{ volts}^2 \text{ per hertz.}$$

Shot Noise and Thermal Noise Combined

The question arises how we calculate the total noise across a resistor due to both the thermal effect and a shot noise current flowing through the resistor.

First of all, we have two different noise processes—one thermal and one shot. It would be ridiculous to assume that these two processes would be in phase. The thermal noise voltage and the shot noise voltage across the resistor added together gives a resultant $V_{nT} = E_{n\,shot} + E_{n\,thermal}$. This formula cannot be applied since the two noise processes are completely unrelated or uncorrelated. The only way we can obtain a resultant is to add the variances of the two noise processes and then extract the square root to obtain the resultant rms voltage. Figure 4-4 illustrates what is meant by a correlated and uncorrelated process. Equation 4-7 can be used to compute the resultant rms noise voltage (E_{nT}) across a resistor due to the shot and thermal effect.

Equation 4-7

$$E_{nT} = \sqrt{4kTRB + 2\,e\,I_{dc}\,B\,R^2}$$

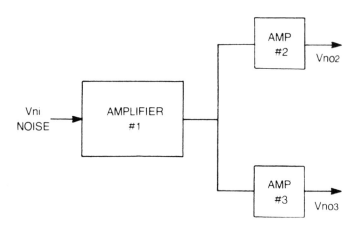

Vno2 AND Vno3 ARE CORRELATED

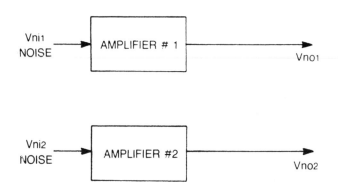

Vno1 AND Vno2 ARE UNCORRELATED

Fig. 4-4. Correlated and uncorrelated processes.

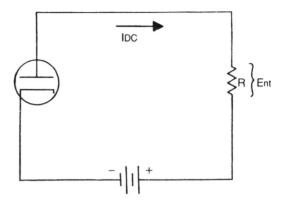

Fig. 4-5. Circuit for Example 4-6.

Refer to Figure 4-5 for a typical circuit illustrating both thermal and shot noise in a diode.

Example 4-6

Refer to Fig. 4-5. The direct current in the circuit is 10 microamps, temperature is 300 degrees Kelvin, the bandwidth is 1 MHz and the resistor is 10 kΩ. Compute the total noise voltage due to the shot and thermal noise processes.

1. Compute the square of the thermal noise voltage across R

$$E_{n\ TH}^2 = 4\ kTRB$$
$$= 4 \times 1.38 \times 10^{-23} \times 3 \times 10^2 \times 10^4 \times 10^6$$
$$\cong 1.66 \times 10^{-10}\ \text{volts}^2$$

2. Compute the square of the shot noise voltage across R.

$$E_{ns}^2 = 2eI_{dc}BR^2$$
$$= 2 \times 1.6 \times 10^{-19} \times 10^{-5} \times 10^6 \times (10^4)^2$$
$$= 3.2 \times 10^{-10}\ \text{volts}^2$$

3. Compute the square of the total noise voltage across R.

$$E_{nT}^2 = 4\ kTRB + 2eI_{dc}BR^2$$

$$= 1.66 \times 10^{-10} + 3.2 \times 10^{-10}$$

$$= 4.86 \times 10^{-10} \text{ volts}^2$$

$$E_{nT} = 2.2 \times 10^{-5}$$

$$= 22 \text{ microvolts.}$$

Now let us look at an extremely unusual problem dealing with both thermal and shot noise. This problem when first stated seems impossible to work since there appears to be insufficient information. However, we will see there is a numerical answer. Example 4-7 states this problem.

Example 4-7

Refer to Fig. 4-6. A resistor produces a thermal noise voltage $E_{n\ TH}$ when it is not connected to any other components (Fig. 4-6A). When the resistor is connected in series with a temperature limited diode as in Fig. 4-6B, the total noise voltage across the resistor becomes 10 times as large as when the resistor was alone. The

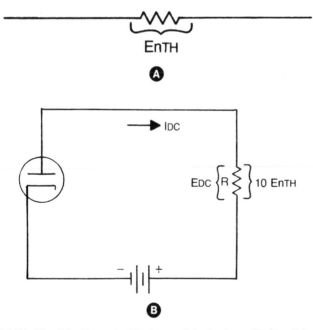

Fig. 4-6.(A). Circuit for Example 4-7 when resistor is alone. (B) Circuit for Example 4-7 when resistor is in series with temperature limited diode.

temperature of the resistor is 300 degrees Kelvin. Compute the average voltage (E_{dc}) across R.

1. The square of the thermal noise voltage across R is:

$$E_n{}^2{}_{TH} = 4kTBR$$

2. When the resistor R is placed in series with the temperature limited diode, the increase in output voltage across R must be due to the shot effect. The new voltage across R is 10 ($E_{n\,TH}$) or 10 $\sqrt{(4kTBR)}$.

In terms of noise voltage squared we have 100 (4kTRB).

This noise voltage squared must be equal to the original thermal noise voltage squared plus the shot noise voltage squared across R. Therefore we may write the following formula:

$$100\ (4kTRB) = 4kTRB + 2eI_{dc}BR^2$$
$$396kTRB = 2eI_{dc}BR^2$$
$$198\ kT = eI_{dc}R.$$

We can observe from the above formula that the product of direct current and resistance is present. This must be the average voltage (E_{dc}).

We may then write:

$$E_{dc} = I_{dc}R = \frac{198\ kT}{e}$$

$$E_{dc} = \frac{198\ kT}{e}$$

As can be seen, the average voltage is proportional to temperature.

Using the value of Boltzmann's constant and the charge on the electron we find that the average voltage across R is:

$$E_{dc} = \frac{198 \times 1.38 \times 10^{-23} \times 3 \times 10^2}{1.6 \times 10^{-19}}$$

$$= 512 \times 10^{-2}$$

$$= 5.12 \text{ volts.}$$

Someone may argue that this voltage is not related to battery voltage since the battery voltage is not present in any of the calculations. However, we must remember that the shot noise is proportional to direct current and that direct current is caused by the battery voltage. The larger the noise output voltage is across R when the diode is added to the circuit, the higher is the shot noise contribution. Since shot noise rises with direct current flow through the diode, an increase in noise voltage across R means that the direct current has increased and there is an increase in E_{dc}. With this in mind, let us work a problem where the noise voltage has increased 100 times from the original noise voltage. We write:

$$10{,}000 \ (4kTRB) = 4kTRB + 2eI_{dc}BR^2$$
$$9999 \ (4kTRB) = 2eI_{dc}BR^2$$
$$19998 \ kTRB = eI_{dc}BR^2$$

$$E_{dc} = \frac{19998 \ kT}{e}$$

$$= \frac{1.9999 \times 10^4 \times 1.38 \times 10^{-23} \times 3 \times 10^2}{1.6 \times 10^{-19}}$$

$$= 517 \text{ volts.}$$

Therefore, an increase in noise voltage across R of 100 times when the diode was switched in resulted in an increase in dc voltage of about 100 times over the original case.

Shot Noise in Relation to Frequency

As we mentioned earlier, shot noise has a white noise spectral density. This means that when shot noise is passed through an amplifier, the only thing that is going to change the spectral density in shape is the frequency response of the amplifier.

If we consider the case of two amplifiers, both of the same bandwidth and gain but of different center frequencies (Fig. 4-7), the shot noise output will be the same for both cases. However, if we allow the bandwidths of each amplifier to be different, then the amplifier with the largest bandwidth will have the highest output voltage.

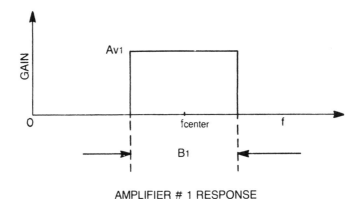

AMPLIFIER # 1 RESPONSE

$A_{v1} = A_{v2}$
$B_1 = B_2$

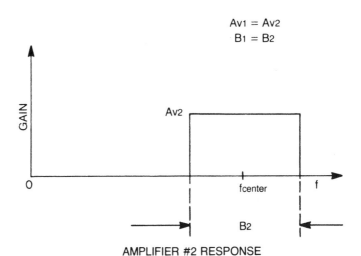

AMPLIFIER #2 RESPONSE

Fig. 4-7. Noise for amplifiers with two equal bandwidths but different frequency regions.

Example 4-8

A shot noise spectral density of 8×10^{-10} volts^2 is applied to two amplifiers which have the same identical response except for center frequency. Each amplifier may be represented as a box response of voltage gain equal to 100.

The first amplifier has extreme frequencies of 80 and 1000 hertz. The second amplifier has extreme frequencies of 300 and 1220 hertz. Compute the shot noise voltage at each amplifier's output.

1. For amplifier one the bandwidth is B $= f_{upper} - f_{lower} =$ 1000 − 80 = 920 hertz.
2. For amplifier two the bandwidth is $B = f_{upper} - f_{lower} =$ 1220 − 300 = 920 hertz.
3. The bandwidth of each amplifier is the same and the gain is the same, also. Each amplifier will have the same output noise voltage. This is determined by the following formula:

$$E_{no}^2 = G \times A_v^2 \times B$$

$$E_{no} = \sqrt{G \times A_v^2 \times B}$$

$$E_{no} = \sqrt{8 \times 10^{-10} \, (10^2)^2 \, (9.2 \times 10^2)}$$

$$= \sqrt{73.6 \times 10^{-4}}$$

$$= 8.58 \times 10^{-2}$$

$$= 85.6 \text{ millivolts.}$$

Another way of looking at this is that the output noise is simply the square root of the area under the output spectral density. This is illustrated in Fig. 4-8.

Shot Noise in PN Junctions

When charge carriers are liberated into potential barrier regions, which occur in pn junctions of semiconductor diodes, and transistors, shot noise occurs. Now, let us examine noise in pn junctions.

The shot noise generated in a pn junction has the same mathematical form as that of the temperature limited vacuum diode. The noise seems to be generated by a noise current generator in parallel with the dynamic resistance of the diode. The equivalent noise circuit for a forward biased pn junction diode is shown in Fig. 4-9. The rms noise current generated is given by Equation 4-8.

Equation 4-8

$$I_{ns} = \sqrt{2 \, e \, I_{dc} \, B}$$

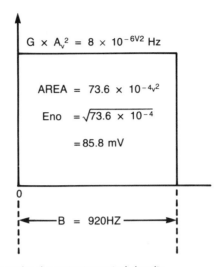

Fig. 4-8. Area under the output spectral density.

Likewise we may determine the shot noise voltage by applying Ohms Law. This voltage is given by Equation 4-9.

Equation 4-9

$$E_{ns} = \sqrt{2eI_{dc}Br_d^2}$$

where r_d is the dynamic resistance of the junction.

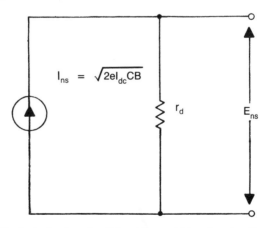

Fig. 4-9. Equivalent noise circuit for a forward biased pn junction.

From electronic physics it is known that the dynamic resistance of a pn junction depends on temperature and the direct current flowing through the junction. The dynamic resistance represents the ratio of a small change in diode voltage to a corresponding change in diode current. Equation 4-10 gives the relation for r_d in terms of temperature and direct current.

Equation 4-10

$$r_d = \frac{kT}{eI_{dc}}$$

To determine the shot noise voltage we may substitute Equation 4-10 into Equation 4-9. The result is Equation 4-11.

Equation 4-11

$$E_{ns} = \sqrt{2kTBr_d}$$

It is obvious from Equation 4-11 that the shot noise voltage across the pn junction has an equation very similar to that of a thermal noise process.

There is a rather interesting relation between the shot noise voltage across the junction and the dynamic resistance r_d. Since r_d is inversely proportional to direct current, the dynamic resistance falls as direct current increases. This causes the shot noise voltage across the junction to decrease. The action seems contrary to the statement we made earlier that shot noise current is proportional to direct current. It would only seem right that if direct current would increase, shot noise voltage would increase. The reason for this strange phenomenon is that shot noise current is proportional to the square root of direct current where dynamic resistance is inversely proportional to direct current. We find that if direct current increases, dynamic resistance falls more quickly than shot noise current rises. The result is that the shot noise voltage becomes inversely related to the direct current. Figure 4-10 illustrates how shot noise junction current varies with direct current.

NOISE IN BIPOLAR TRANSISTORS

In a bipolar transistor there are three main types of noise that

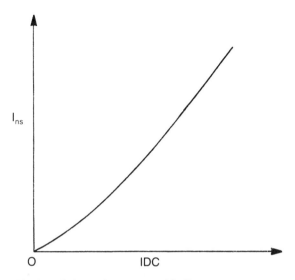

Fig. 4-10. Varying of shot noise current with direct current.

are generated. These types are thermal, shot and flicker noises. Thermal noise is caused by thermal agitation of electrons in the base spreading resistance of the transistor. Shot noise is the result of two mechanisms—noise due to recombination of electrons and holes in the base region and noise from random carrier motion across the emitter and collector junctions.

The flicker effect results from fluctuations in carrier density due to various surface leakage effects. The spectrum of this noise varies inversely with frequency. This flicker noise is a "pink noise process." The fluctuations in carrier density cause the conductivity of the semiconductor material to vary, resulting in varying noise voltage when direct current is flowing through the junction. Flicker noise is most serious at low frequencies since its spectral density is highest at low frequencies. Figure 4-11 shows a low frequency model for a transistor noise circuit. We note the following:

1. E_{nb} represents the thermal noise voltage given by Equation 4-12.

Equation 4-12

$$E_{nb} = \sqrt{4\ k\ T\ r_b B}$$

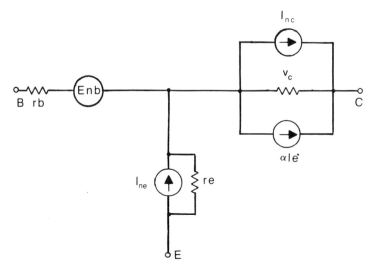

Fig. 4-11. Low frequency model for a transistor noise circuit.

2. I_{nc} and I_{ne} represent shot noise current processes related to collector and emitter respectively. Their equations are 4-13 and 4-14.

Equation 4-13

$$I_{nc} = \sqrt{2eIcB}$$

Equation 4-14

$$I_{ne} = \sqrt{2eI_eB}$$

In a transistor it is typical for the overall frequency behavior for the spectral density to look like that of Fig. 4-12. We note that at low frequencies the flicker effect plays a large role in the spectral density. The density drops at the rate of 3 dB per octave until it becomes constant at some low frequency, usually lower or in the vicinity of 1 kHz. At higher frequencies, the density begins to rise up again at 20 dB per decade, starting at a frequency near

$$f_\alpha \sqrt{1 - \alpha_0},$$

where fα is the alpha cutoff frequency of the transistor and αo is the short circuit current gain in the common base configuration at low frequencies.

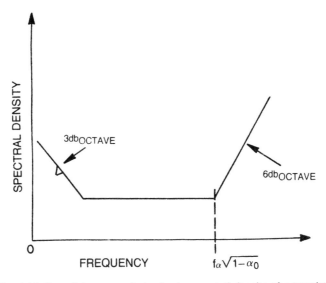

Fig. 4-12. Overall frequency behavior for spectral density of a transistor.

Example 4-9

A transistor has a low frequency alpha of 0.91 and an alpha cutoff frequency of 100 kHz. Where in the frequency spectrum will the transistor noise begin to rise at about 20 dB per decade?

The frequency will be $f = f\alpha \sqrt{1 - \alpha_o}$.

$$f = 100 \text{ kHz} \sqrt{1 - .91}$$
$$= 100 \text{ kHz} \sqrt{.09}$$
$$= 100 \text{ kHz} (.3)$$
$$= 30 \text{ kHz.}$$

We will find later in this book that the noise increase at higher frequencies can cause problems when evaluating the overall behavior of a network.

OTHER FORMS OF NOISE IN TRANSISTORS

Whenever current must divide between two or more paths, the result is random fluctuations in the division. The result is partition

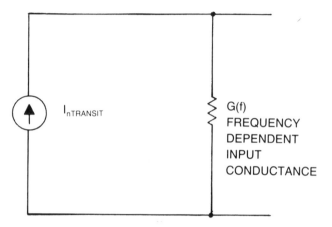

Fig. 4-13. Example of transit time or high frequency noise equivalent circuit.

noise which has a flat spectrum. Therefore we consider this process "white noise."

In a transistor there takes a finite amount of time for a charge carrier to cross the base-emitter or base-collector junction. When this time is comparable to the period of the input signal of the transistor circuit, some of the charge carriers diffuse back to their origin. The result is that the input admittance of the transistor has a frequency dependent conductive part. As frequency rises this conductance also rises, producing more noise current and allowing the spectral density to rise. Figure 4-13 shows an example of a transit time or high frequency noise equivalent circuit.

NOISE IN FIELD-EFFECT TRANSISTORS

The noise in a field effect transistor is mainly due to the thermal noise of the conducting channel. Figure 4-14 shows an equivalent noise model for a field effect transistor. We note that a resistor has been shown in the gate lead which represents the noise generating mechanism in the transistor.

Another form of noise in transistors is known as "popcorn noise." This type of noise appears as a "burst" and results when there is a change in input bias current. The noise density is strongest at the lower frequencies usually below several hundred hertz. The cause of popcorn noise is believed to be due to surface imperfections in the semiconductors.

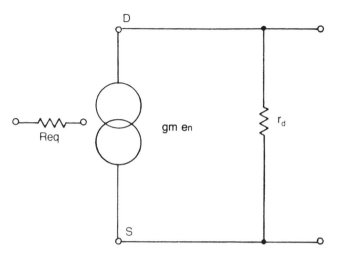

Fig. 4-14. Equivalent noise model for a field effect transistor.

NOISE IN VACUUM TUBES

Vacuum tubes generate noise due mainly to the shot effect. However, with tubes having more than one positive element, partition noise occurs.

EQUIVALENT NOISE RESISTANCE

Although the noise produced by an electronic device or even a complete system may be due to many different types of noise, we may simplify the calculation of noise at the output by using an equivalent noise resistance which represents a resistor (fictitious) that produces the noise.

Refer to Fig. 4-15A. The system is noisy and produces a noise output V_{no}. In Fig. 4-15B, the system is thought of as being noiseless and producing the same output noise voltage, since a noisy resistor R has been connected across the input terminals. This resistor is producing a thermal noise and is representing the noise generating properties of the system. We must remember that this resistor is physically not present in the circuit. The resistance must not be in a position to draw any dc.

The noise equivalent resistance has been determined for many electronic devices. The most familiar is the Req for field effect transistors and vacuum tubes.

The noise equivalent resistance of a triode is approximately that given by Equation 4-15.

Equation 4-15

$$R_{eq} = \frac{2.5}{gm}$$

where gm is the transconductance of the vacuum tube.

For a pentode where the screen grid is at a positive potential, extra noise is generated which is due to the partition effect. The noise equivalent resistance becomes that given by Equation 4-16.

Equation 4-16

$$R_{eq} = \left(\frac{2.5}{gm} + \frac{20\,I_{c2}}{gm^2} \right) \frac{I_b}{I_b + I_{c2}}$$

where I_{c2} is the screen current and I_b is the plate current.

Example 4-10

Compute the equivalent noise resistance of a pentode if *gm* =

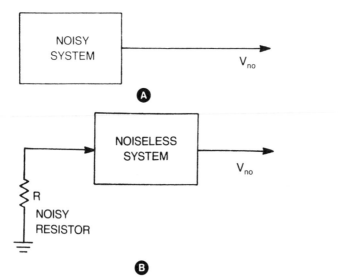

Fig. 4-15(A). Noisy system. (B) Noiseless system with noisy resistor R.

5000 $\mu\Omega$, I_b = 7.5 ma, and I_{c2} = 4.5 ma.

$$R_{eq} = \left(\frac{2.5}{gm} + \frac{20\,I_{c2}}{gm^2} \right) \left(\frac{I_b}{I_b + I_{c2}} \right)$$

$$= \left(\frac{2.5}{5 \times 10^{-3}} + \frac{20 \times 4.5 \times 10^{-3}}{(5 \times 10^{-3})^2} \right) \left(\frac{7.5}{7.5 + 4.5} \right)$$

$$= 2563\ \Omega$$

Vacuum tubes having several positive elements like a pentagrid converter may have very large noise equivalent resistances such as several hundred thousand ohms.

For a field-effect transistor (FET) the noise equivalent resistance is given by a formula that is nearly the same as that for the triode. The R_{eq} for an FET is given by Equation 4-17.

Equation 4-17

$$R_{eq} = \frac{\lambda}{gm}$$

where λ is a number normally between .2 and .8.

It must be remembered that noise equivalent resistance is stated for a particular operating condition.

Noise Equivalent Resistance of a System

Consider the circuit shown in Fig. 4-16A. The amplifier has a certain noise equivalent resistance which represents the noise producing property of the amplifier. To what is this resistance due? It is related to the noise equivalent resistance of each amplifying device and each resistance in the amplifier. Consider the amplifier shown in Fig. 4-16B. There is an input resistance R_i, an output load resistor R_L, and a noise equivalent resistance of the amplifier device itself. To determine the noise equivalent resistance of the entire amplifier, we must first reflect R_L back to the input circuit. To do this we remember that R_L is producing a thermal noise of $E_n^2 = 4kTRB$. To reflect this back we divide E_n^2 by the square of the voltage gain of the amplifier. This is shown in Fig. 4-16B and is given by $\frac{R_L}{A_v^2}$. The total noise producing resistance of the

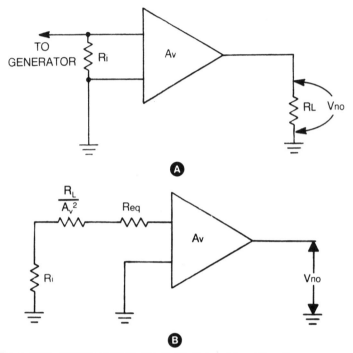

Fig. 4-16(A). Typical amplifier circuit. (B) Determining "Reg$_r$".

amplifier will be given by Equation 4-18.

Equation 4-18

$$R_{eqT} = R_i + R_{eq} + \frac{R_L}{A_v^2}$$

The output noise of the amplifier is given in Equation 4-19.

Equation 4-19

$$E_{no} = \sqrt{4kT (R_i + R_{eq} + R_L) \frac{B(A_v)^2}{A_v^2}}$$

Suppose a generator of internal resistance Rs was connected to the circuit as shown. What would be the new noise output voltage? The generator's internal resistance Rs is in parallel with Ri. Thus the equivalent noise resistance of the entire circuit becomes that given by Equation 4-20.

Equation 4-20

$$R_{eqT} = \frac{R_i R_s}{R_i + R_s} + R_{eq} + \frac{R_L}{A_v^2}$$

The output noise is now given by Equation 4-21.

Equation 4-21

$$V_{no} = \sqrt{4kT \left[\frac{(R_i R_s}{R_i + R_s} + R_{eq} + \frac{R_L)}{A_v^2} \right] B(A_v)}$$

The input signal of the amplifier is that of Equation 4-22.

Equation 4-22

$$V_i = V_s \frac{R_i}{R_i + R_s}$$

Note that the equivalent noise resistance had nothing to do with calculating the input noise voltage. This should seem obvious since R_{eq} is not physically present in the circuit. Therefore it cannot help determine the input signal voltage. Also, the reflection of R_L is not of any importance since this resistor is not physically present. Only the actual resistive components R_i and R_s determine the amount of input voltage of the amplifier.

Noise Equivalent Resistance of Stages in Cascade

Refer to Fig. 4-17. We can see several amplifier stages in cascade. The noise equivalent resistances are given as R_{eq1}, R_{eq2} ... R_{eqn}. We know that when a resistance is reflected from output to input of an amplifier, the resistance is diminished by a value equal to the square of the voltage gain. When a resistance is reflected from the output circuit of an amplifier backwards through two stages of amplification, the resistance is diminished by the product of the squares of the individual voltage gains. Therefore, the equivalent noise resistance of the total amplifier combination will

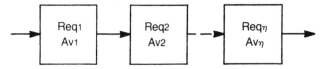

Fig. 4-17. Determining noise equivalent resistance for several stages in cascade.

be that given by Equation 4-23.

Equation 4-23

$$R_{eqT} = R_{eq1} + \frac{R_{eq2}}{A_{v1}^{2}} + \frac{R_{eq3}}{(A_{v1}^{3})(A_{v2}^{2})} + \frac{----- R_{eqn}}{(A_{v1})^{2}(A_{v2})^{2}---(A_{vn-1})^{2}}$$

Example 4-11

An amplifier has an input resistance of 50 kΩ and FET with a noise equivalent resistance of 14 kΩ and an output load resistor of 10 kΩ. The gain of the amplifier is 10. Compute the total noise equivalent resistance.

$$R_{eqt} = R_i + R_{eq} + \frac{R_l}{A_v^{2}}$$

$$= 50 \text{ k}\Omega + 14 \text{ k}\Omega + \frac{10 \text{ k}\Omega}{(10)^{2}}$$

$$= 50 \text{ k}\Omega + 14 \text{ k}\Omega + 0.1 \text{ k}\Omega$$

$$= 64.1 \text{ k}\Omega.$$

In this example the contribution to the total noise equivalent resistance was quite small due to the effect of gain in the formula.

Example 4-12

Three amplifier stages connected in cascade have the following data:

Amplifier	Voltage Gain	R_{eq}
#1	20	5 kΩ
#2	10	10 kΩ
#3	2	100 kΩ

Compute the overall noise equivalent resistance of the total amplifier.

$$R_{eqT} = R_{eq1} + \frac{R_{eq2}}{A_{v1\;2}} - \frac{R_{eq3}}{A_{v1}^{2}\,A_{v2}^{2}}$$

$$= 5\,k\Omega \;+\; \frac{10\;k\Omega}{(20)^2} \;+\; \frac{100\;k\Omega}{(20)^2\,(10)^2}$$

$$= 5\;k\Omega \;+\; 25\;\Omega \;+\; 2.5\;\Omega$$

$$= 5,027.5\;\Omega$$

Why didn't we use the voltage gain for stage three in our calculation? The noise equivalent resistance of stage three is present at the input of the amplifier because of our definition of this resistance.

Example 4-13

Repeat problem 4-12 if Amplifier #1 has a gain of unity such as in a follower circuit.

$$R_{eqT} = 5\;k\Omega \;+\; 10\;k\Omega \;+\; \frac{100\;k\Omega}{(1)^2\,(10)^?}$$

$$= 5\;k\Omega \;+\; 10\;k\Omega \;+\; 1\;k\Omega$$

$$= 16\;k\Omega$$

We see that since the gain of the follower circuit is unity, the noise equivalent resistance becomes larger, in this case three times larger than when the first amplifier stage had a voltage gain of 20.

SIGNAL-TO-NOISE RATIO

Signal-to-noise ratio is the ratio of signal power to noise power or the ratio of signal voltage to noise voltage, existing at some particular point in an electronic system. It is more common to determine signal to noise ratio in terms of power ratios.

Why Is Signal-To-Noise Ratio Used?

The signal-to-noise ratio gives us information about the signal and noise strength at a particular point in a circuit. Assuming a power ratio, suppose the signal to noise gives a quotient of five. This means that the signal power is five times as much as the noise power. Signal to noise should normally be as large as possible to ensure a good quality output signal. Some signals are capable of withstanding high levels of noise and still being useful.

Expression of Signal-To-Noise Ratio

We may write the formula for signal-to-noise ratio as shown in Equation 4-24.

Equation 4-24

$$\frac{S}{N} = \frac{P_s}{P_n}$$

where S and P_s are the signal power, and N and P_n are the noise power.

If P_s and P_n dissipate power in resistor R then we may write the following: $P_s = e_s^2/R$ and $P_n = e_n^2/R$ where e_s and e_n are the signal and noise voltage respectively. We can then write the formula for signal-to-noise ratio in terms of signal and noise voltages as follows in Equation 4-25.

Equation 4-25

$$\frac{S}{N} = \frac{P_s}{P_n} = \frac{\dfrac{e_s^2}{R}}{\dfrac{e_n^2}{R}} = \frac{e_s^2}{e_n^2}$$

$$\frac{S}{N} = \frac{e_s^2}{e_n}$$

We can thus conclude that the resistance value is not important as long as we know the signal and noise voltages across the resistor. We will learn later in Chapter 6 how signal-to-noise-ratio is related to a figure of merit we call the noise figure.

Signal-to-Noise Ratio in dB

We may write signal-to-noise ratio in decibels by using Equation 4-26.

Equation 4-26

$$\frac{S}{N}\bigg|_{dB} = 10 \; log \; \frac{S}{N}$$

where S/N is the signal-to-noise power ratio. The limits on $S/N \, dB$ are negative infinity dB and positive infinity dB.

SUMMARY

1. Two extremely important types of noise in electronic devices are the shot effect and partition noise. Both types of noise are related to a direct current.

2. Shot effect noise is dependent on both bandwidth and direct current.

3. Shot and thermal noise voltages may be combined to give a resultant providing we take the square root of the sum of their squares.

4. Shot noise spectral density for practical purposes is considered independent on frequency.

5. The equivalent noise resistance of a device describes the degree of noisiness of the device.

6. Signal-to-noise ratio describes the relative size of signal and noise power at a particular point in the circuit.

PROBLEMS AND QUESTIONS

1. Compute the square of the rms noise current flowing through a temperature limited diode if the direct current is 0.1 ampere and the bandwidth is 30 kHz.

2. Suppose the diode mentioned in Problem 1 has a resistor of 4 kΩ in series with it. Compute the rms noise voltage across R due to the shot effect, thermal effect, and both effects combined ($T = 290°$ K).

3. Compute the spectral densities for Problem 2 in amps^2 per hertz and volts^2 per hertz.

4. A shot noise spectral density of 5×10^{-10} volts^2 per hertz is applied to the input of an amplifier with a voltage gain of 40 and bandwidth of 1 kHz. Assuming a box shaped frequency response, calculate the output noise voltage.

5. Compute the equivalent resistance of a triode if the transconductance is 4000 μV.

6. An amplifier has an input resistance of 10 kΩ, an electronic device with equivalent noise resistance of 2 kΩ and an output load resistance of 40 kΩ. The voltage gain of the amplifier is five. Compute the noise equivalent resistance for the system.

7. Four amplifier stages connected in cascade have the following data:

Amplifier	Voltage Gain	R_{eq}
#1	5	50 kΩ
#2	10	100 kΩ
#3	20	200 kΩ
#4	5	30 kΩ

Compute the total noise equivalent resistance.

8. Refer to Fig. 4-17. Derive an expression for the signal to noise power ratio for the circuit if the signal and noise contributions are those caused only by the source.

9. The equivalent noise resistance of an amplifier is 4 kΩ and the voltage gain is 50. Compute the output noise voltage of the amplifier. The temperature is 69 degrees Fahrenheit. Bandwidth = 10 kHz.

10. When analyzing an amplifier circuit, why must the noise equivalent resistance be shown in a position where it draws no current?

11. List the types of noise present in a bipolar transistor and give the reason for their generation.

12. Compute the shot noise voltage across a pn junction if the temperature is 20 degrees Centigrade, bandwidth is 4 kHz and the dynamic resistance is 25 Ω.

13. Compute the signal-to-noise ratio in dB if the signal to noise ratio is 12.

14. Sketch a spectral density diagram for a bipolar transistor, showing the flicker effect region and the region where noise spectral density begins to rise at higher frequencies.

Chapter 5

Noise Calculations
in Advanced Networks

S O FAR WE HAVE EXAMINED THE EFFECT OF NOISE IN NET-works composed of resistors only. The question arises, how do we make noise calculations in networks composed of capacitors, inductors and resistors as well as amplifiers? We now will address this problem starting first from the problem of passing white noise through an ideal amplifier.

WHITE NOISE PASSED THROUGH AN IDEAL AMPLIFIER

When a thermal noise or shot noise spectral density is present at the input of an ideal amplifier (one which has constant gain over its bandwidth), the output of the amplifier will have a spectral density which is equal to the input spectral density given in volts squared per Hertz multiplied by the square of the voltage gain. See Fig. 5-1 for this situation.

The rms noise voltage is then determined by taking the square root of the product of the output noise spectral density and the bandwidth.

It is important to remember that the area under the spectral density as shown in Fig. 5-1 represents the square of the rms noise voltage.

The condition that has been discussed here deals with a "theoretical situation," the ideal amplifier. Now we will begin to look

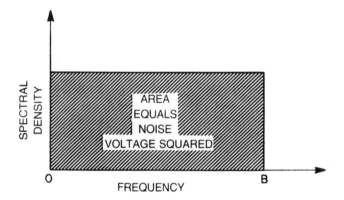

Fig. 5-1. The area under the spectral density plot can represent the square of the rms noise voltage.

at practical situations where circuits have frequency dependent characteristics that influence the amount of output noise.

Example 5-1

A resistor R is generating thermal noise. Determine the output noise voltage of the ideal amplifier that has this resistor connected to the input terminals. The voltage gain is Av, and the bandwidth is B.

1. The spectral density of the resistor is:

$$G_{(f)} = 4kTR$$

2. The spectral density at the output of the amplifier is:

$$A_v^2 G_{(f)} = A_v^2 (4kTR)$$

3. The output rms noise voltage is then

$$E_{no} = \sqrt{4kTRA_v^2(B)}$$

or

$$E_{no} = (\sqrt{4kTRB}) \, Av.$$

The solution of this problem simply states mathematically that the rms noise output voltage is equal to the product of voltage gain and the rms noise voltage of the resistor.

DO INDUCTORS AND CAPACITORS GENERATE NOISE?

Inductive and capacitive reactance do not generate any thermal noise. This only seems correct since a reactance cannot dissipate power. Refer to Fig. 5-2. Suppose the inductor has a reactance that delivers a noise power to the resistor. Also, suppose the resistor delivers noise power to the reactance. If thermal equilibrium exists, the reactance must receive power equal to the power delivered by the resistance. Since the dissipated power of the reactance is the received power and an inductive reactance cannot dissipate power, then the power delivered by the reactance must be zero. The same ideas apply to a capacitor having a capacitive reactance.

Thermal noise can be generated in an inductor's resistive part. A physical inductor has both inductance and resistance. This resistance can generate thermal noise since it is the same as the process associated with a physical resistor. A capacitor may produce a small thermal noise associated with its lead resistance and leakage resistance. Figure 5-3 summarizes the previous statements.

WHAT EFFECT DO INDUCTORS AND CAPACITORS PLAY IN DETERMINING NOISE

The reactive properties of inductors and capacitors work in conjunction with circuit resistance in determining the limiting bandwidth for a noise process. When dealing with thermal noise, in circuits composed of resistors and reactive components, the noise seems to be generated by the real part of the circuit impedance.

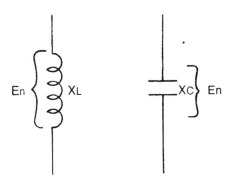

Fig. 5-2. Reactances do not generate thermal noise.

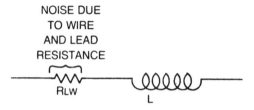

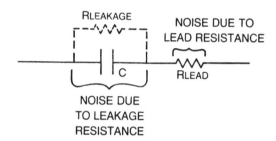

Fig. 5-3. Thermal noise may be generated in the resistive parts of capacitors and inductors.

The real part of a circuit's impedance will be related to the inductive and capacitive reactances in the circuit. The real part will vary with frequency and the rms noise voltage squared will be given by taking the noise contributions at all frequencies. The only way this can be done is by calculus. The formula is given below as Equation 5-1.

Equation 5-1

$$E_n^2 = \int_0^\infty 4kTR_{e(f)}df$$

where R_e is the real part of the circuit impedance.

Example 5-2

Determine the real part of the impedance of the following circuits shown in Fig. 5-4.

1. For circuit A in Fig. 5-4, the impedance is given by $Z = R + j\omega L$

where R represents the real part of the circuit impedance.
2. For circuit B in Fig. 5-4, the impedance is given by

$$Z = \frac{R\,(j\omega\,L)}{R + j\omega L}$$

$$= \frac{(Rj\omega L)\,(R - j\omega L)}{(R + j\omega L)\,(R - j\omega L)}$$

$$= \frac{R^2 j\omega L}{R^2 + (\omega L)^2} + \frac{\omega^2 L^2 R}{R^2 + (\omega L)^2}$$

The real part of the impedance in this example is

$$\frac{R\,\omega^2 L^2}{R^2 + (\omega L)^2}$$

THERMAL NOISE IN OUTPUT
OF PARALLEL TUNED CIRCUIT

When you are confronted with determining the thermal noise at the output terminals of a tuned circuit, you must ask the follow-

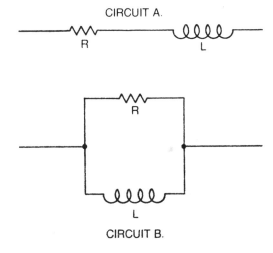

CIRCUIT A.

CIRCUIT B.

Fig. 5-4. Circuits for Example 5-2.

ing question. How does the real part of the impedance behave over the bandwidth of interest?

For example, if you are interested in a very small bandwidth surrounding the resonant frequency of the tuned circuit, you may assume the real part of the impedance is constant over the bandwidth of interest. No calculus is necessary to do this problem.

However, if the real part varies considerably over the bandwidth of interest, then a calculus solution is necessary using Equation 5-1. An example would be the case where you were asked to determine the overall rms noise voltage for the circuit due to frequency contributions from zero to infinity.

Example 5-3

The tuned circuit shown in Fig. 5-5 is to have its output noise voltage evaluated within a frequency range of 10 kHz centered around the center frequency.

1. Since the circuit is unloaded, the circuit magnification factor is determined as follows:

$$Q = \frac{1}{R}\sqrt{\frac{L}{C}} \qquad\qquad Q = \frac{1}{10}\sqrt{\frac{1\times 10^{-6}}{1\times 10^{-12}}}$$

$$Q = \frac{1}{10\sqrt{10^{6}}} \qquad\qquad Q = \frac{1}{10}(1000) = 100$$

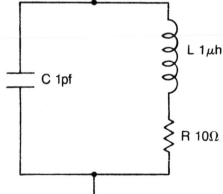

Fig. 5-5. Circuits for Example 5-3.

2. With a Q of 100 we may calculate the center frequency of the circuit using the following formula which is valid for $Q > 10$.

$$f_c = \frac{1}{2\pi\sqrt{L\,C}} = \frac{.159}{\sqrt{1 \times 10^{-6} \times 1 \times 10^{-12}}}$$

$$= \frac{.159}{\sqrt{10^{-18}}} = .159 \times 10^9 = 159 \text{ MHz}$$

3. Calculate the 3 dB bandwidth

$$BW = \frac{fc}{Q} = \frac{1.59 \times 10^8}{1 \times 10^2} = 15.9 \times 10^5$$

$$BW = 1.59 \times 10^6 = 1.59 \text{ MHZ}$$

4. Since the 3 dB bandwidth is much larger than the bandwidth in which we want to determine the noise, we may assume that the real part of the circuit impedance remains constant over the frequency range in question.

5. The real part may be taken to be equal to the impedance at the center frequency which is approximately,

$$Z = \frac{L}{CR} = \frac{1 \times 10^{-6}}{1 \times 10^{-12} \times 1 \times 10} = R_r$$

$$= \frac{10^{-6}}{10^{-11}} = 10^5 \Omega$$

6. Calculate the noise voltage E_n

$$E_n^2 = 4kTR_r B$$

$$= 4 \times 1.38 \times 10^{-23} \times 2.9 \times 10^2 \times 10^5 \times 10^4$$

$$= 1.6 \times 10^{-11} \text{ volts}^2$$

$$E_n = \sqrt{16.0 \times 10^{-12}} = 4.0 \times 10^{-6}$$

$$E_n = 4.0 \mu V.$$

NOISE EQUIVALENT BANDWIDTH

Earlier in this chapter we observed the long procedure for computing the noise at the output terminals of a network, which was made of both resistive and reactive components. As was observed, we were required to go to calculus for the determination of the noise. Now we will try to understand the concept of noise equivalent bandwidth. This bandwidth will speed up the calculation of total noise output since tables of noise equivalent bandwidth are available in many engineering handbooks.

To appreciate the topic of noise equivalent bandwidth consider the simple RC network shown in Fig. 5-6. Suppose you are asked to determine the noise voltage across the circuit terminals for a bandwidth of zero to infinity. The procedure will be as follows using Equation 5-1,

$$E_n^2 = 4kT \int_0^\infty Re(f)df$$

where $Re(f)$ is the real part of the circuit impedance. The real part of the circuit impedance is found in the following manner

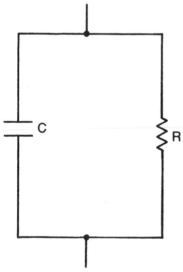

Fig. 5-6. Basic RC network.

$$Z = \frac{R\,(1/j\omega C)}{R + 1/j\omega C} = \frac{R}{j\omega CR + 1}$$

Conjugating,

$$Z = \frac{R(-j\omega CR + 1)}{(j\omega CR + 1)\,(-j\omega CR + 1)} = \frac{-j\omega CR^2 + R}{(\omega CR^2)\quad 1}$$

$$Z = \frac{R}{1 + (\omega CR)^2} - j\,\frac{\omega CR^2}{1 + (\omega CR)^2}$$

where $\dfrac{R}{1 + (\omega CR)^2}$ is the real part of the impedance and

$\dfrac{-j\omega CR^2}{1 + (\omega CR)^2}$ is the imaginary part of the circuit impedance. Placing the real part into the equation we obtain

$$E_n^{\,2} = 4kT \int_0^\infty \frac{R\,df}{1 + (\omega CR)^2}$$

Preparing the integral we obtain

$$E_n^{\,2} = \frac{4kT}{2\pi CR} \int_0^\infty \frac{R\,df\,2\pi CR}{1 + (\omega CR)^2} = \frac{2kTR}{\pi CR} \int_0^\infty \frac{dx}{1 + x^2}$$

The integral $\displaystyle\int_0^\infty \frac{dx}{1 + x^2}$ is of the form $\displaystyle\int_0^\infty \frac{dx}{a^2 + x^2} = \frac{1}{a} \arctan \frac{x}{a}$

where a equals one.

$$E_n^{\,2} = \frac{2kT}{\pi C} \arctan [x] \int_0^\infty$$

$$x = \omega CR$$

133

$$E_n^2 = \frac{2kT}{\pi C} [\arctan \infty - \arctan 0]$$

$$E_n^2 = \frac{2kT}{\pi C} [\frac{\pi}{2} - 0]$$

$$E_n^2 = \frac{kT}{C}$$

Equation 5-2

$$E_n^2 = kT/C$$

We see from Equation 5-2 that the total rms noise voltage is proportional to Boltzmanns constant and temperature and inversely proportional to capacitance. This of course was proven in Chapter 2.

Now let us use the concept of noise equivalent bandwidth:

1. We look up the noise equivalent bandwidth for the circuit in an engineering handbook and find it to be:

$$B_{eg} = \frac{\pi}{2} fc$$

2. We know that the cutoff frequency for the network is:

$$f_c = \frac{1}{2 \pi RC}$$

3. We then write:

$E_n^2 = 4kTBR$ where B is in this case the noise equivalent bandwidth.

$$E_n^2 = 4kT (1/2\pi RC) (\pi/2) R = kT/C.$$

Notice that the answer is the same as that using the previous method. Even though we have not shown how to find noise equivalent bandwidth at this time, we can now appreciate the ease with making noise calculations.

What is Noise Equivalent Bandwidth?

When working a problem where you were to find the thermal noise in a resistor for a specified bandwidth, you were actually assuming that an ideal filter of response, equal to unity and bandwidth equal to the specified bandwidth, was sampling the resistor's spectral density. To understand this, let us say that a resistor produces a special density of 6×10^{-14} volt ^{2}/Hz. The bandwidth is 10 kHz.

The spectral density of the noise looks like that in Fig. 5-7. Since the bandwidth was to be 10^4 Hz, there seems to be a filter present that is selecting the noise in a bandwidth of 10 kHz. This is shown in Fig. 5-8.

Notice how easy the calculation is if the filter is ideal. However, we know very well that such things are theoretical. All of the circuits we are confronted with have some dependence on frequency. If we could convert all of the frequency dependent circuits we deal with into a frequency independent circuit like the ideal filter, all of our calculations would be simple. In effect, this is just what the concept of noise equivalent bandwidth does.

Look at the following two responses of two different circuits, shown in Fig. 5-9. Both of these circuits could have an equivalent ideal filter response.

To determine the noise equivalent bandwidth of a particular network, we must find the equivalent ideal filter bandwidth for that network. We learned earlier that the square of noise voltage is related to the area of the response in question. If we were to equate

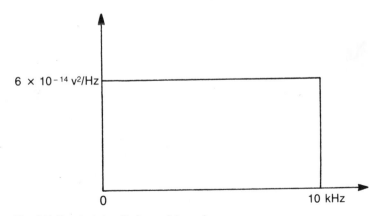

Fig. 5-7. Spectral density for resistor noise.

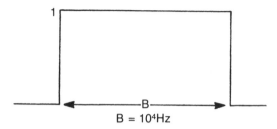

Fig. 5-8. Effective filter response when assigning a bandwidth to a resistor.

the area under the actual response with the area of an ideal low pass filter, we could then determine the equivalent bandwidth for the circuit in question.

In formula form we have Equation 5-3.

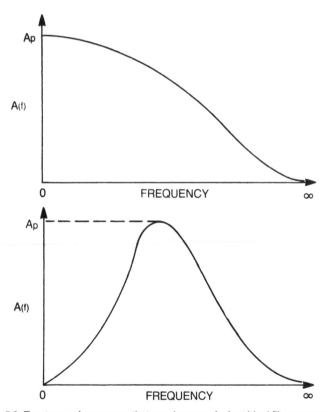

Fig. 5-9. Two types of responses that may have equivalent ideal filter response.

Equation 5-3

NOISE EQUIVALENT
BANDWIDTH OF
CIRCUIT IN   =
QUESTION

AREA UNDER SQUARE OF CIRCUIT
VOLTAGE RESPONSE
————————————————————————————
SQUARE OF CIRCUIT VOLTAGE RESPONSE PEAK

(If dealing with a bandpass type response use the response at midband. If dealing with a low-pass response use the response at zero frequency for the peak).

Or,

$$B_{eq} = \frac{\int_0^\infty |A(f)|^2\, df}{A_p^2}$$

where $A(f)$ is voltage response of circuit in question as a function of frequency. Symbol A_p is the peak voltage response. If the circuit in question is a bandpass type, A_p is just the voltage gain or transfer function at midband frequency. Let us now work an example using noise equivalent bandwidth.

Example 5-4

We have a source of white noise connected at the input of a low pass filter. The source of white noise has a spectral density of 4×10^{-4} volt^2.

First we determine the noise equivalent bandwidth.

Let us say that the low pass filter is made of an amplifier with voltage gain equal to 20 followed by RC network of the form shown in Fig. 5-10. We will assume that the noise produced by the resistor R is insignificant in respect to the noise applied to the circuit.

The response of our circuit is as follows:

$$\frac{E_{out}}{E_{in}} = \frac{20}{1 + jwCR}$$

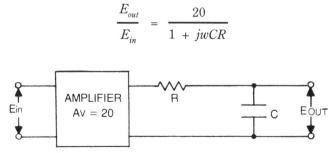

Fig. 5-10. Circuit for Example 5-4.

The response at zero frequency is:

$$\frac{E_{in}}{E_{out}} = 20$$

Applying the formula for noise equivalent bandwidth we have:

$$B_{eq} = \frac{\displaystyle\int_0^\infty \left|\frac{20}{1 + j\omega CR}\right|^2 df}{(20)^2}$$

$$= \frac{\displaystyle\int_0^\infty \frac{400}{1 + (\omega CR)^2} df}{400}$$

$$= \int_0^\infty \frac{df}{1 + (\omega CR)^2}$$

$$= \int_0^\infty \frac{df}{1 + (\omega CR)^2} = \frac{1}{2\pi CR}\int_0^\infty \frac{df \times 2\pi CR}{1 + (\omega CR)^2}$$

$$= \frac{1}{2\pi CR} \times \frac{\pi}{2} = f_c \times \frac{\pi}{2}$$

Our result says that the noise equivalent bandwidth is equal to $\pi/2$ times the cutoff frequency of the amplifier circuit.

To compute the total output noise we use the formula:

$$E_{no}^2 = G_{(f)} \times A_p^2 \times B_{eq}$$

$$= 4 \times 10^{-4} \times (20)^2 \times \pi/2 \times fc.$$

Our problem now is to find the cutoff frequency. Since the filter part of the amplifier was made of a 1 kΩ resistor and 1 μf capacitor, we may compute the cutoff frequency by applying the formula

$$f_c = \frac{1}{2\pi RC}$$

$$f_c = \frac{1}{2\pi \times 1 \times 10^3 \times 1 \times 10^{-6}} = \frac{.159}{10^{-3}}$$

$$= 159 \text{ Hz}$$

Using the cutoff frequency in our formula,

$$E_{no}^2 = 4 \times 10^{-4} \times (400) \times 1.57 \times 1.59 \times 10^2$$

$$E_{no}^2 = 40 \text{ volts}^2$$

$$E_{no} = 6.32 \text{ volts.}$$

Example 5-5

Determine the noise equivalent bandwidth of the circuit shown in Fig. 5-11.

Write the transfer function for the circuit.

$$\frac{E_o}{E_i} = \frac{\frac{1}{j\omega C}}{R + j\omega L + 1/j\omega C}$$

$$\left| \frac{E_o}{E_i} \right| = \frac{1}{\sqrt{(1 - \omega^2 LC)^2 + (\omega CR)^2}}$$

$$= \frac{1}{\sqrt{1 - 2\omega^2 LC + \omega^4 L^2 C^2 + \omega^2 C^2 R^2}}$$

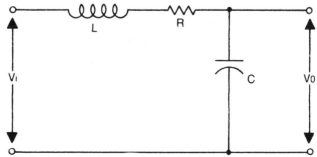

Fig. 5-11. Circuit for Example 5-5.

$$= \frac{1}{\sqrt{1 - 2\omega^2 L C + \omega^2 C^2 R^2 + \omega^4 L^2 C^2}}.$$

This transfer function is somewhat difficult to work with when evaluating noise equivalent bandwidth. By examination of the function we can see that the term $\omega^4 L^2 C^2$ would be the determining factor in the frequency response. We will make a rough estimate of the response as being

$$\frac{E_o}{E_i} = \frac{1}{\sqrt{1 + \omega^4 L^2 C^2}}$$

To determine the response maximum, we note that the response is low pass. We allow the frequency to be zero and we obtain,

$$A_p = \frac{1}{1 + 0} = 1$$

Applying the formula for noise equivalent bandwidth, we obtain

$$B_{eq} = \frac{\displaystyle\int_0^\infty \frac{1}{1 + \omega^4 L^2 C^2} \, df}{1^2}$$

$$= \int_0^\infty \frac{df}{1 + \omega^4 L^2 C^2}$$

How do we evaluate such an integral? The answer is use the following formula:

$$\int_0^\infty \frac{X^{m-1} \, dx}{1 + X^n} = \frac{\pi}{n} \bigg/ \sin \frac{m\pi}{n}$$

By comparison of our formula for B_{eq} with this integral, we note the following:

$$x = \omega\sqrt{LC}$$
$$n = 4$$

also, $m = 1$

Preparing the integral,

$$B_{eq} = \frac{1}{2\pi\sqrt{L\,C}} \int_0^\infty \frac{2\pi LCdf}{1 + (\omega\sqrt{LC})^4}$$

$$= fr \int_0^\infty \frac{dx}{1 + x^4}$$

$$B_{eq} = \frac{1}{2\pi\sqrt{LC}}\left(\frac{\pi}{4} \middle/ \sin \frac{1(\pi)}{4}\right)$$

$$= fr \;\; \frac{\pi}{4} \middle/ \sin \frac{\pi}{4}$$

$$= fr \;\; \frac{\pi}{4} \middle/ \frac{1}{\sqrt{2}}$$

$$B_{eq} = fr \;\; \frac{\pi}{4} \cdot \frac{\sqrt{2}}{1} = \frac{fr\,\pi}{2\sqrt{2}}$$

This states that the noise equivalent bandwidth is equal to the resonant frequency of the network multiplied by

$$\frac{\pi}{2\sqrt{2}}$$

SUMMARY

1. Thermal noise appears to be generated by the real part of a circuit's impedance.

2. The real part of a circuit's impedance is not the circuit's resistor value alone.

3. When dealing with tuned circuits the real part of the circuit impedance may be taken as the maximum impedance value which occurs at resonance, providing that the frequency range of interest is spread about the circuit's center frequency, and the frequency spread is much smaller than the actual 3 dB bandwidth of the tuned circuit.

4. Noise equivalent bandwidth enables us to make noise calculations quickly, since tables of noise equivalent bandwidth for various types of circuits have been evaluated.

PROBLEMS AND QUESTIONS

1. Compute the real part of the impedances for the circuits shown in Fig. 5-12.

2. For the circuit shown in Fig. 5-13 calculate rms noise output voltage if the input spectral density is 5×10^{-8} volts²/Hz.

3. Compute the noise equivalent bandwidth for the circuit shown in Fig. 5-14. Give answer in terms of R and C.

4. Make a plot of the real part of the circuit impedance versus frequency for the circuit in Fig. 5-14.

5. Discuss what is meant by noise equivalent bandwidth.

6. Determine the noise equivalent bandwidth of a filter that has a transfer function of

$$H(jw) = e^{-0.35 \left(\frac{f}{B}\right)^2}$$

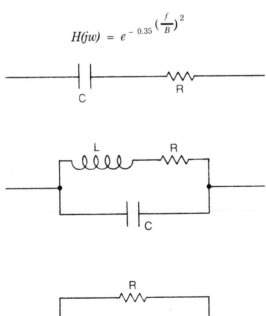

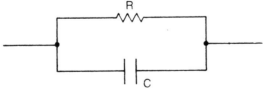

Fig. 5-12. Circuits for Problem 1.

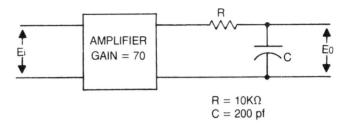

R = 10KΩ
C = 200 pf

Fig. 5-13. Circuit for Problem 2.

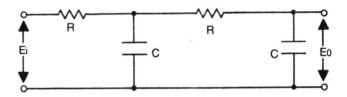

Fig. 5-14. Circuit for Problem 3.

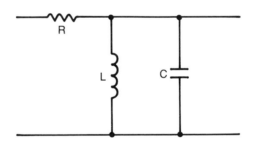

Fig. 5-15. Circuit for problem 5-10.

7. Set up the integral to determine the thermal noise associated with a circuit composed of a resistor of resistance R and an inductor of inductance L in parallel.

8. Given the following circuits:

   a. 10 kΩ in parallel with 1 nF
   b. 100 kΩ in parallel with 1 nF.

Is there any difference in the total noise for each circuit? Why?

What is the noise output for each circuit in a bandwidth of zero to 10 kHz?

143

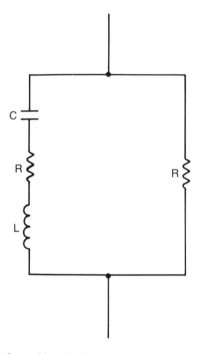

Fig. 5-16. Circuit for problem 5-11.

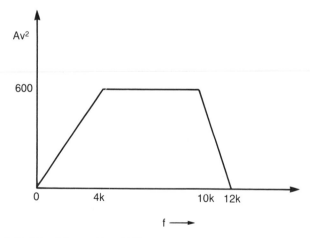

Fig. 5-17. Circuit for problem 5-13.

9. What are the noise equivalent bandwidths of each circuit in problem 8?

10. What is the noise equivalent bandwidth for the circuit of Fig. 5-15.

11. What is the real part of the impedance of the circuit in Fig. 5-16?

12. What is the noise equivalent bandwidth of a circuit with a transfer function of

$$H_{(f)} = 10 \, e^{-0.35 \left(\frac{f}{B} \right)}$$

13. What is the noise equivalent bandwidth of the function shown in Fig. 5-17.

14. Find the noise equivalent bandwidth for a circuit that has the transfer function of

$$H_{(f)} = 10e^{-\frac{f}{B}}$$

Chapter 6

Noise Figure
and Noise Temperature

N OISE FIGURE IS A MEASURE OF THE AMOUNT OF DEGRADING
of signal-to-noise ratio from input to output of a system. Noise
temperature is an alternate means of expressing noise figure since
it rates the noise generating section of the system in terms of an
equivalent noise temperature instead of a degrading factor.

NOISE FIGURE

In formula form, noise figure is the quotient of dividing signal-
to-noise ratio at the input of a system by the signal-to-noise ratio
at the output of the same system. Stated mathematically we have
Equation 6-1.

Equation 6-1

$$F = \frac{\dfrac{S_i}{N_i}}{\dfrac{S_o}{N_o}}$$

where:

S_i = signal power at input of system

146

N_i = noise power at input of system

S_o = signal power at output of system

N_o = noise power at output of system

We can take Equation 6-1 and restate it as:

$$F = \frac{\dfrac{S_i}{N_i}}{\dfrac{S_o}{N_o}} = \frac{S_i}{N_i} \times \frac{N_o}{S_o}$$

Equation 6-2

$$\text{Thus, } F = \frac{S_i}{S_o} \times \frac{N_o}{N_i}$$

Now the power gain of the system is stated as $G = S_o/S_i$ where G is the power gain.

We can then write Equation 6-3.

Equation 6-3

$$F = \frac{1}{G} \times \frac{N_o}{N_i}$$

The minimum output noise is simply the input noise multiplied by the power gain or $N_i(G)$. However, in a practical situation the noise output always exceeds the input noise multiplied by the gain. The amount of extra noise is N_n. Thus, we can write the output noise as:

Equation 6-4

$$N_o = N_i(G) + N_n$$

The noise figure may then be written as:

Equation 6-5

$$F = \frac{N_o}{N_i(G)} = \frac{N_i(G) + N_n}{N_i(G)}$$

or,

Equation 6-6

$$F = 1 + N_n/N_i(G)$$

As can be seen from Equation 6-6, if N_n were zero (an ideal situation) then the noise figure would be unity. Thus, the ideal system situation in regards to noise figure would be a noise figure of unity.

As an example, a noise figure of six would mean that the system would be adding five times as much noise in the output as the output contribution, due to the product of power gain and input noise power, ($N_n = 5\ N_iG$).

Likewise, a noise figure of 10 would mean that the system was producing nine times as much noise in the output as the product of power gain and input noise. ($N_n = 9\ N_iG$)

Example 6-1

The input signal-to-noise ratio of a system is 30 and the output signal-to-noise ratio is four. Compute the noise figure.

$$F = \frac{\dfrac{S_i}{N_i}}{\dfrac{S_o}{N_o}} = 30/4 = 7.5$$

Example 6-2

The extra noise supplied by a system in the output is 50 microwatts. The noise in is five microwatts and the power gain is four. What is the output noise and the noise figure?

1. $N_o = N_i(G) + N_n = 5\ \mu W\ (4) + 50\ \mu W = 70\ \mu W$

$$2. \quad F = 1 + \frac{N_n}{N_i \, (G)} = 1 + \frac{50 \; \mu W}{5 \; \mu W \, (4)} = 3.5.$$

Example 6-3

In Example 6-2 what is the ratio of extra noise to the minimum output noise?

Minimum output noise would be: $N_i \, (G) = 5 \; \mu W(4) = 20 \; \mu W$. The ratio would be:

$$\frac{N_n}{N_i \, (G)} = \frac{50 \; \mu W}{20 \; \mu W} = 2.5.$$

Example 6-4

The noise figure of a system is 13. What is the output signal-to-noise ratio if the input signal-to-noise ratio is 65?

$$\frac{S_o}{N_o} = \frac{\dfrac{S_i}{N_i}}{F} = \frac{65}{13} = 5.$$

Noise Figure in Decibels

The noise figure of a system may be determined in decibels. The formula is given by Equation 6-7.

Equation 6-7

$$F_{dB} = 10 \log F$$

The ideal noise figure in decibels is determined by the case when F = unity. Thus,

$$F_{dB} = 10 \log 1 = 0 \; dB$$

Example 6-5

What is the noise figure in dB if F is 20?

$$F_{dB} = 10 \log 20 = 13 \; dB.$$

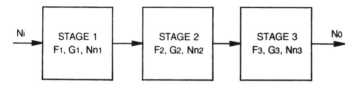

Fig. 6-1. Example of several stages connected in cascade.

Noise Figure of Several Stages in Cascade

Suppose we have several stages connected in cascade as shown in Fig. 6-1.

Where F_1, F_2, F_3, are the respective noise figures of the individual stages,

G_1, G_2, G_3, are the respective power gains,

N_{n1}, N_{n2}, N_{n3}, are the respective extra noise component powers of each stages.

By our basic formula for noise figure we have:

$F = N_o/N_i$ (G) where the output is:

$No = N_i$ (G_1) (G_2) (G_3) $+ N_{n1}$ (G_2) (G_3) $+ N_{n2}$ (G_3) $+ N_{n3}$

If we divide this by the minimum possible output noise which is N_i (G_1) (G_2) (G_3) we obtain,

Equation 6-8

$$F = \frac{N_i\,G_1\,G_2\,G_3 + N_{n1}\,G_2\,G_3 + N_{n2}\,G_3 + N_{n3}}{N_i\,G_1\,G_2\,G_3}$$

which reduces to:

$$1 + \frac{N_{n1}}{N_i\,G_1} + \frac{N_{n2}}{N_i\,G_1\,G_2} + \frac{N_{n3}}{N_i\,G_1\,G_2\,G_3}$$

Now F_1 would be the noise figure for the first stage. It could be written as follows: $F_1 = 1 + N_{n1}/N_i\,G_1$, likewise

$$F_2 = 1 + \frac{N_{n2}}{N_i\,G_2} \quad \text{and} \quad F_3 = 1 + \frac{N_{n3}}{N_i\,G_3}$$

We can rearrange the formulas for F_2 and F_3 to become:

$$F_2 - 1 = \frac{N_{n2}}{N_i \, G_2} \qquad\qquad F_3 - 1 = \frac{N_{n3}}{N_1 G_3}$$

Therefore, Equation 6-8 can be rewritten as:

Equation 6-9

$$F = \frac{F_1 + F_2 - 1}{G_1} + \frac{F_3 - 1}{G_1 \, G_2}$$

In more general terms we could state that the noise figure for a cascaded stage would be:

Equation 6-10

$$F = \frac{F_1 + F_2 - 1}{G_1} + \frac{F_3 - 1}{G_1 \, G_2} + \cdots + \frac{F_n - 1}{G_1 \, G_2 \cdots G_{n-1}}$$

It can be seen that if the gains are larger than one, then the noise figure is highly dependent on the first stages since all of the terms except the first one are diminished by a gain constant.

Example 6-6

A two stage amplifier has a noise figure and power gain of 10 and 5 respectively for the first stage. If the noise figure for the second stage is 7 and its power gain is 14, find the overall noise figure.

$$F = F_1 + \frac{F_2 - 1}{G_1}$$

$$= 10 + \frac{7 - 1}{5}$$

$$= 11.2$$

Notice that the power gain of the second stage had nothing to do with the calculation of the overall noise figure.

Example 6-7

A two stage amplifier with cascade connection has the following data:

	Stage 1	Stage 2
noise figure	3dB	10dB
power gain	4	5

Compute the overall noise figure.

First you must compute the noise figure as a pure number for each stage:

$$FdB_1 = 10 \ log \ F_1 \qquad\qquad FdB_2 = 10 \ log \ F_2$$
$$3 = 10 \ log \ F_1 \qquad\qquad\quad 10 = 10 \ log \ F_2$$
$$3/10 = log \ F_1 \qquad\qquad 10/10 = log \ F_2$$
$$0.3 = log \ F_1 \qquad\qquad\quad 1 = log \ F_2$$
$$F_1 = 2 \qquad\qquad\qquad\quad F_2 = 10$$

Then we can apply the formula for two stages in cascade:

$$F = F_1 + (F_2 - 1)/G_1$$
$$F = 2 + (10 - 1)/4$$
$$F = 2 + 9/4$$
$$F = 4.25.$$

NOISE FIGURE FOR PARALLEL CONNECTED STAGES

Suppose we have several stages connected in parallel as shown in Fig. 6-2.

Noise figures are F_1 and F_2.

Extra noise components are N_{n1} and N_{n2}.

Power gains are G_1 and G_2.

From block diagram algebra, it can be shown that the overall gain of two amplifiers in parallel and not interacting becomes the sum of the original gains of the amplifier. For example, the two amplifiers of Fig. 6-2 produce an overall power gain of $G_1 + G_2$. The minimum output noise of the circuit is simply the input noise power multiplied by the total power gain. This becomes $N_i [G_1 +$

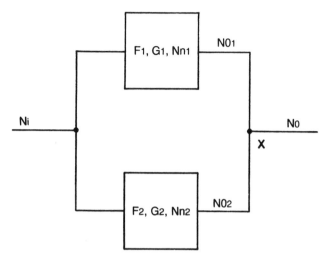

Fig. 6-2. Several stages connected in parallel.

G_2]. The respective noise outputs for amplifier one and two are listed below:

Amplifier 1	Amplifier 2
$N_{o1} = N_i (G_1) + N_{n1}$	$N_{o2} = N_i (G_2) + N_{n2}$

The total output noise is the sum of the two output noise values.

$$N_o = N_{o1} + N_{o2} = N_i G_1 + N_{n1} + N_i G_2 + N_{n2}$$

The overall noise figure can then be written as

$$F = \frac{N_o}{N_i G} = \frac{N_i G_1 + N_{n1} + N_i G_2 + N_{n2}}{N_i (G_1 + G_2)}$$

$$F = \frac{N_i (G_1 + G_2) + N_{n1} + N_{n2}}{N_i (G_1 + G_2)}$$

Equation 6-11

$$F = 1 + \frac{N_{n1} + N_{n2}}{(G_1 + G_2) N_i}$$

For n stages in parallel, the noise figure becomes:

Equation 6-12

$$F = 1 + \frac{N_{n1} + N_{n2} + \ldots\ldots\ldots + N_{nn}}{(G_1 + G_2 + \ldots\ldots + G_n)\, N_i}$$

Equation 6-12 is not very convenient when determining noise figure unless we know the actual values of the various noise powers. We can rearrange this formula by remembering that:

$$F_1 - 1 = N_{n1}/N_i\, G_1 \qquad\qquad F_2 - 1 = N_{n2}/N_i\, G_2$$

and thus,

$$G_1\, (F_1 - 1)\, N_i = N_{n1} \qquad\qquad G_2\, (F_2 - 1)\, N_i = N_{n2}$$

For a two stage parallel network:

$$N_n = G_1(F_1 - 1)N_i + G_2(F_2 - 1)\, N_i$$

$$F = 1 + \frac{N_{n1} + N_{n2}}{(G_1 + G_2)\, N_i}$$

$$F = 1 + \frac{G_1\, (F_1 - 1)\, N_i + G_2\, (F_2 - 1)\, N_i}{(G_1 + G_2)\, N_i} = \frac{G_1\, F_1 + G_2\, F_2}{G_1 + G_2}$$

Example 6-8

You are given a four stage amplifier, each stage in parallel. The respective gains are: $G_1 = 10$, $G_2 = 15$, $G_3 = 10$, $G_4 = 5$, and the respective extra noise components are $N_{n1} = 10\ \mu W$, $N_{n2} = 20\ \mu W$, $N_{n3} = 40\ \mu W$, and $N_{n4} = 15\ \mu W$. The noise input is $2\ \mu W$. What is the noise figure?

$$F = 1 + \frac{(10\mu W + 20\mu W + 40\mu W + 15\mu W)}{(10 + 15 + 10 + 5)\, 2\mu W}$$

$$F = 1 + \frac{85 \ \mu W}{80 \ \mu W} \cong 2.1$$

NOISE FIGURE FOR PARALLEL AND SERIES NETWORKS

Examine Fig. 6-3A and notice that we have a circuit composed of both parallel connected networks and a series connected network. The noise at point x is $N = N_i (G_1 + G_2) + N_{n1} + N_{n2}$ and all of that noise becomes amplified by gain G_3 and added to N_{n3}. The total noise output is:

$$N_o = (N_i (G_1 + G_2) + N_{n1} + N_{n2}) G_3 + N_{n3}$$

The minimum output noise that could occur would be $N_i (G_1 + G_2) G_3$.

Thus,

$$F = N_o/N_i G = \frac{(N_i (G_1 + G_2) + N_{n1} + N_{n2}) G_3 + N_{n3}}{N_i (G_1 + G_2) G_3}$$

or,

$$F = \frac{N_i (G_1 + G_2) G_3 + (N_{n1} + N_{n2}) G_3 + N_{n3}}{N_i (G_1 + G_2) G_3}$$

Equation 6-13

$$F = 1 + \frac{(N_{n1} + N_{n2})}{N_i(G_1 + G_2)} + \frac{N_n}{N_i(G_1 + G_2) G_3}$$

$$F = F_{12} + \frac{F_3 - 1}{G_1 + G_2}$$

Example 6-9

Given the following network shown in Fig. 6-3B, what is the overall noise figure?

We know that the noise at the output of network #1 is $N = N_i G_1 + N_{n1}$, and that the noise at the output of the two parallel networks would be the product of the noise going into those net-

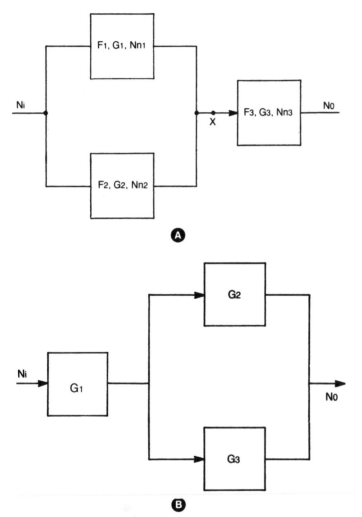

Fig. 6-3(A). Stages in parallel and series. (B) System for Example 6-9.

works multiplied by the gain of the parallel combination of networks two and three. The minimum output noise that could occur would be $N_i (G_1) (G_2 + G_3)$

From the definition of noise figure we know:

$$F = \frac{S_i/N_i}{S_o/N_o} = \frac{N_o}{N_i G} = \frac{(N_i G_1 + N_{n1}) (G_2 + G_3) + N_{n2} + N_{n3}}{N_i \, G_1 \, (G_2 + G_3)}$$

$$F = \frac{N_i\,G_1\,(G_2 + G_3)}{N_i\,G_1\,(G_2 + G_3)} + \frac{N_{n1}\,(G_2 + G_3) + N_{n2} + N_{n3}}{N_i\,G_1\,(G_2 + G_3)}$$

$$F = 1 + \frac{N_{n1}}{N_i\,G_1} + \frac{N_{n2} + N_{n3}}{N_i\,(G_2 + G_3)\,G_1}$$

Equation 6-14

$$F = F_1 + \frac{F_{23} - 1}{G_1}$$

As can be seen, the noise figure is highly dependent on F_1 provided $G_1 > 1$.

We have just looked at a few of the possible noise figure problems that can occur. Now let us look at some miscellaneous problems.

Example 6-10

Refer to Fig. 6-4. If the overall noise figure is six, what is the gain of the network #2 and its extra noise component N_{n2}?

1. $F = N_o/(N_iG)$
2. $G = N_o/(N_iF) = 3.33$
3. $G = G_1\,G_2$ therefore, $G_2 = G/G_1 = 3.33/3 = 1.11$
4. $F = F_1 + (F_2 - 1)/G_1$ where
   $F_1 = 1 + N_{n1}/(N_i\,G)$
   $F_1 = 1 + 6\,\mu W(2(3)\mu W) = 2$

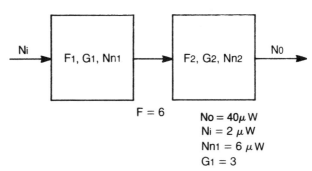

Fig. 6-4. Problem for Example 6-10.

Now $F = 6$ thus,

$$6 = 2 + (F_2 - 1)/G_1, \ 4 = (F_2 - 1)/3$$

$$F_2 = 13$$

$$\frac{N_{n2}}{N_i G_2} = F_2 - 1, \ \frac{N_{n2}}{2(1.11)} = 12$$

$$N_{n2} = 2 \times 12 \ (1.11) \cong 26.6 \ \mu W$$

COMPUTING NOISE FIGURE FROM PRACTICAL CIRCUITS

Let us now learn how noise figure can be calculated for some practical circuits.

Example 6-11

Compute the noise figure for the circuit in Fig. 6-5.

1. Find input signal-to-noise ratio

$$V_i = (V_S) \left(\frac{R_1 + R_2}{R_1 + R_2 + R_s} \right)$$

$$E_{ni} = \sqrt{(4kTRsB)} \left(\frac{R_1 + R_2}{R_1 + R_2 + R_s} \right)$$

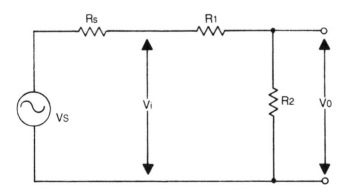

Fig. 6-5. Circuit for Example 6-11.

$$S_i = V_s^2 \frac{\left(\dfrac{R_1 + R_2}{R_1 + R_2 + R_s}\right)^2}{R_1 + R_2}$$

$$N_i = \frac{(4kTRsB)\left(\dfrac{R_1 + R_2}{R_1 + R_2 + R_s}\right)^2}{R_1 + R_2}$$

$$\frac{S_i}{N_i} = \frac{\dfrac{V_S^2\left(\dfrac{R_1 + R_2}{R_1 + R_2 + R_s}\right)^2}{R_1 + R_2}}{\dfrac{(4kTRsB)\left(\dfrac{R_1 + R_2}{R_1 + R_2 + R_s}\right)^2}{R_1 + R_2}}$$

$$\frac{S_i}{N_i} = \frac{V_s^2}{4kTRsB}$$

2. Find output signal-to-noise ratio.

$$V_o = (V_s) \left(\frac{R_2}{R_1 + R_2 + R_s}\right)$$

$$E_{no} = \sqrt{4kTRTB}$$

where

$$R_T = \frac{(R_s + R_1)\,R_2}{R_s + R_1 + R_2}$$

$$S_o = (V_s^2) \left(\frac{\dfrac{R_2^2}{(R_1 + R_2 + R_s)^2}}{R_2}\right)$$

$$N_o = \frac{4kTR_TB}{R_2}$$

$$S_o = \frac{(V_s^2)(R_2)}{(R_1 + R_2 + R_s)^2}$$

$$= V_S^2 \times \frac{R^2}{(R_1 + R_2 + R_3)^2} \times \frac{R_2}{4kTR_TB}$$

$$= V_s^2 \; \frac{R_2^2}{(R_1 + R_2 + R_s)^2 \; \dfrac{(4kT \times R_2 (R_1 + R_s)B)}{R_1 + R_2 + R_s}}$$

$$= \frac{V_s^2 \, R_2^2}{(R_1 + R_2 + R_s)(4kTR_2(R_1 + R_s)B)}$$

$$F = \frac{\dfrac{S_i}{N_i}}{\dfrac{S_o}{N_o}} = \frac{\dfrac{V_S^2}{4kTR_sB}}{\dfrac{V_s^2 R_2^2}{(R_1 + R_2 + R_s)(4k \, T \, R_2 \, (R_1 + R_s) \, B)}}$$

$$= \frac{(R_1 + R_2 + R_s)(R_1 + R_s)}{R_2 \, R_s}$$

$$= \frac{R_1^2 + R_1R_s + R_1R_2 + R_2R_s + R_1R_s + R_s^2}{R_2R_s}$$

$$F = \frac{R_1^2}{R_2R_s} + \frac{R_1}{R_s} + \frac{R_1}{R_2} + 1 + \frac{R_s}{R_2}$$

$$F = 1 + \frac{2R_1}{R_2} + \frac{R_1}{R_s} + \frac{R_s}{R_2} + \frac{R_1^2}{R_2 \, R_s}$$

Example 6-11B

For circuit 6-5 calculate F if:

$R_1 = 10 \text{ k}\Omega$

$R_2 = 10 \text{ k}\Omega$

$R_s = 5 \text{ k}\Omega$

$$F = 1 + 2 \times \frac{10\ k}{10\ k} + \frac{10\ k}{5\ k} + \frac{5\ k}{10\ k} + \frac{(10\ k)^2}{(10\ k)(5\ k)}$$

$$= 1 + 2 + 2 + .5 + 2$$

$$= 7.5$$

Example 6-11C

Compute noise figure of Fig. 6-6.

$$1. \quad V_i = V_s \times R_L\,(R_s + R_L)$$

$$E_{ni}^2 = 4kTR_sB \times (R_L/(R_s + R_L))^2$$

$$S_i = V_s^2 R_L^2/(R_s + R)^2/R_L$$

$$N_i = (4kTR_sB/R_L)/(R_L(R_s + R_L))^2$$

$$\frac{S_i}{N_i} = \frac{V_s^2\ \dfrac{R_L^2}{(R_S + R_L)^2\ R_L}}{\dfrac{4kTR_sB}{R_L}\ \dfrac{(R_l)^2}{(R_S + R_l)^2}}$$

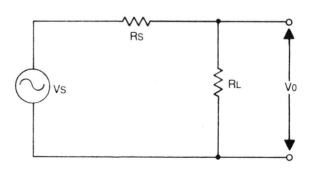

Fig. 6-6. Circuit for Example 6-13.

$$\frac{S_i}{N_i} = \frac{V_s^2}{4kTR_sB}$$

2. $V_o = V_s \times R_L/R_s + R_l$

   $E_{no} = 4kT \, ((R_sR_L)/(R_s + R_L))B$

   $P_o = V_s^2/R_L \times R_L^2/(R_s + R)^2$

$$N_o = 4kT \, \frac{\dfrac{R_s \, R_L \, B}{R_s + R_L}}{R_L}$$

$$\frac{S_o}{N_o} = \frac{\dfrac{V_s^2}{R_L} \times \dfrac{R_l^2}{(R_s + R_L)^2}}{\dfrac{4 \, kT}{R_L} \times \dfrac{R_sR_lB}{R_s + R_L}}$$

$$= \frac{V_s^2}{4kTB} \times \frac{R_L^2}{(R_s + R_L)^2} \times \frac{R_s + R_L}{R_sR_L}$$

$$\frac{V_s^2}{4kTBR_s} \times \frac{R_L}{R_s + R_L}$$

$$F = \frac{\dfrac{S_i}{N_i}}{\dfrac{S_o}{N_o}} = \frac{\dfrac{V_s^2}{4kTR_sB}}{\dfrac{V_s^2}{4kTRB_s} \times \dfrac{R_l}{R_s + R_L}}$$

$$= \frac{R_s + R_L}{R_l}$$

$$F = 1 + R_s/R_L.$$

Example 6-11D

For the circuit in Fig. 6-6 compute the noise Figure if $R_L = R_s$.

$$F = 1 + R_s/R_L$$

$$F = 1 + 1 = 2$$

$$F_{dB} = 10 \log 2 = 3 \; dB.$$

Example 6-12A

An amplifier has a noise equivalent resistance of 400 Ω and is connected to a generator as shown in Fig. 6-7. What is the noise figure if $R_s = 100 \; \Omega$?

$$F = 1 + \frac{N_n}{N_i G} = 1 + \frac{\dfrac{N_n}{G}}{N_i}$$

$$= 1 + \frac{4kTR_{eq}B}{4kTR_s B} = 1 + \frac{R_{eq}}{R_s}$$

$$= 1 + \frac{400}{100} = 5.$$

Example 6-12B

What is the minimum noise figure for Example 6-12A if we vary R_s?

$$F = 1 \text{ if } R_s = \infty$$

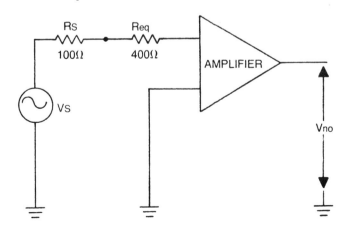

Fig. 6-7. Circuit for Example 6-11A.

If we vary R_{eq} then

$$F = 1 \text{ if } R_{eq} = 0.$$

NOISE EQUIVALENT TEMPERATURE

Consider the formula for noise figure that we developed earlier. The formula was:

$$F = 1 + N_n/GN_i$$

Suppose we divide N_n and GN_i by G. Then $F = 1 + \dfrac{N_n}{GN_i}$

In other words we are looking at both noise sources at the input. Now consider that both noises were caused by thermal effects (we know that these sources are really many different types of noises combined). Each noise N_n and N_i may be considered to be caused by a resistor. Let each resistor be different. The resistor that produces N_n/G may be considered R_{eq}, the thermal equivalent resistance of the amplifier, and the resistor producing N_i some other value R_s. Refer to Fig. 6-8.

Now let $R_{eq} = R_s$. The only thing we can do to adjust the formula for proper value is to define new temperatures T_{eq} and T_o.

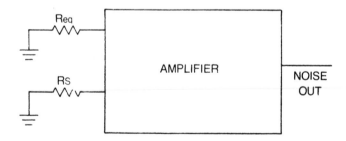

LET Req = Rs
THEN Req IS AT Teq
AND Rs IS AT To

Fig. 6-8. Determining noise equivalent temperature.

Equation 6-15

$$F = 1 + \dfrac{\dfrac{N_n}{G}}{N} = 1 + \dfrac{R_{eq}}{R_s} = 1 + \dfrac{T_{eq}}{T_o}$$

Thus we have a new formula for noise figure which is:

$$F = 1 + T_{eq}/T_o$$

where T_{eq} = noise equivalent temperature

T_o = Reference temperature
(Source temperature)

Noise equivalent temperature is thus written as follows:

Equation 6-16

$$T_{eq} = T_o (F - 1)$$

Various reference temperatures have been used including 290, 293 and 300 degrees Kelvin. These temperatures correspond to 17, 20, and 27 degrees Centrigrade respectively. We always state the temperature in degrees Kelvin.

We must remember in all of these discussions that noise equivalent temperature is not a physical temperature and cannot be measured with a thermometer. It is merely a means of describing the noisiness of a circuit.

To obtain a feeling for the relation between noise figure and noise temperature, assume a reference temperature of 290 degrees Kelvin and vary F. Calculate T_{eq}.

F	$T_{eq} = T_o (F - 1)$
1	0
1.005	1.45
1.05	14.5
1.5	145
2	290
3	580
4	870
5	1160

F	$T_{eq} = T_o(F - 1)$
6	1450
10	2610
20	5510

We notice that small noise figures give fairly large temperatures. Noise equivalent temperature is preferred in many instances since a noise equivalent temperature, that is a value say 1.4 degrees, is easier to understand than a noise figure of 1.005.

Example 6-13

Compute the noise equivalent temperature if $F = 13$ and $T_o = 290$ degrees.

$$T_{eq} = (T_o)(F - 1) = 290(13 - 1)$$
$$= 290(12)$$
$$= 3480\,°k$$

RELATION BETWEEN NOISE EQUIVALENT RESISTANCE AND NOISE EQUIVALENT TEMPERATURE

We know: $F = 1 + R_{eq}/R_s$
and $F = 1 + T_{eq}/T_o$

If we equate both of the above formulas, $1 + R_{eq}/R_s = 1 + T_{eq}/T_o$ solving for R_{eq}/R_s:

Equation 6-16

$$\frac{R_{eq}}{R_S} = \frac{T_{eq}}{T_o}$$

This formula says that the ratio of noise equivalent resistance to source resistance is the same as the ratio of noise equivalent temperature to reference temperature.

Example 6-14

An FET with a noise equivalent resistance of 500 Ω is connected in an amplifier with a generator source resistance of 250 Ω. On the basis of a reference temperature of 290 degrees Kelvin, what is the noise equivalent temperature?

$$\frac{R_{eq}}{R_s} = \frac{T_{eq}}{T_0}$$

$$T_{eq} = T_0 \frac{(R_{eq})}{R_s} = 290 \, (500/250)$$

$$= \quad 580 \text{ degrees Kelvin.}$$

Example 6-15

What is the noise equivalent temperature and noise equivalent resistance if a certain noisy device is connected to a source resistance of 1000 Ω, with $T_0 = 290$ degrees and the noise figure is 8.

$$\frac{R_{eq}}{R_s} = \frac{T_{eq}}{T_0} = F - 1$$

$$\frac{R_{eq}}{R_s} = F - 1, \quad \frac{R_{eq}}{1000} = 8 - 1 = 7, \, R_{eq} = 7000\Omega$$

$$\frac{T_{eq}}{T_0} = F - 1, \quad \frac{T_{eq}}{290} = 8 - 1 = 7,$$

$$T_{eq} = 2030 \text{ degrees Kelvin.}$$

Example 6-16

A parametric amplifier has a noise figure of 1.0004. The reference temperature is 290. Compute noise equivalent temperature and the ratio of R_{eq} to R_s.

$$\frac{T_{eq}}{T_0} = F - 1, \quad \frac{T_{eq}}{290} = 1.0004 - 1$$

$$\frac{T_{eq}}{290} = 4 \times 10^{-4}$$

$$T_{eq} = .1160 \text{ degrees Kelvin}$$

$$\frac{R_{eq}}{R_s} = \frac{T_{eq}}{T_o}$$

$$= \frac{.1160}{290}$$

$$= 4 \times 10^{-4}.$$

Example 6-17

For what noise figure is the noise equivalent temperature and noise reference temperature the same?

$$F = 1 + T_{eq}/T_o$$
$$\text{If } T_{eq} = T_o, T_{eq}/T_o = 1$$
$$F = 1 + 1 = 2.$$

Example 6-18

A network has an input noise of 5 μW and a power gain of seven. If the noise reference temperature is 290 degrees Kelvin, what is the network noise and the noise figure if T_{eq} = 2900 degrees Kelvin?

$$F = 1 + N_n/N_iG, F - 1 = N_n/N_iG$$
$$F - 1 = N_n/(5 \ \mu\text{W}) \ (7)$$
$$35 \ \mu\text{W} \ (F - 1) = N_n.$$

$$F - 1 = \frac{N_n}{N_iG} = \frac{T_{eq}}{T_o}$$

$$F - 1 = \frac{N_n}{N_iG} = \frac{2900}{290} = 10, F = 11$$

$$N_n = N_iG \ (F - 1) = (35 \ \mu\text{W}) \ (11 - 1)$$

$$= (35 \ \mu W) \ (10) = 350 \ \mu W$$

Noise temperature may be extremely small for maser amplifiers and may be unusually large for gaseous lamps. It is not uncommon for some type of lamps to have noise figures in excess of 40 which correspond to noise equivalent temperatures of nearly 12,000 °K.

What is the equivalent noise temperature of a resistor? This is just the temperature that the resistor must be at to produce the output noise if we know the bandwidth.

NOISE EQUIVALENT TEMPERATURE DEFINITION

We can represent the source resistance by a temperature T_o, and allow the noise equivalent resistance R_{eq} and source resistance R_s to be the same. Then the temperature to which the noise equivalent resistance must be raised, to result in the same amount of noise output as when the source and noise equivalent temperature were equal and their resistive values unequal, would be called the "noise equivalent temperature."

NOISE EQUIVALENT TEMPERATURE FOR SEVERAL STAGES IN CASCADE

If several stages are in cascade we can define the over-all noise equivalent temperature by using the formula for noise figure which is

$$F = F_1 + \frac{F_2 - 1}{G_1} + \cdots + \frac{F_n - 1}{G_1 G_2 \ldots G_{n-1}}$$

We know $T_e = T_o \ (F - 1)$

$$T_e = T_o \left(F_1 + \frac{F_2 - 1}{G_1} + \cdots + \frac{F_n - 1}{G_1 G_2 \cdot G_n - 1} \right)$$

$$= T_o \left(F_1 + \frac{F_2 - 1}{G_1} + \cdots + \frac{F_n - 1}{G_1 G_2 \ldots G_{n-1}} - 1 \right)$$

$$= T_o \ (F_1 - 1) + \frac{T_o \ (F_2 - 1)}{G_1} + \ldots + \frac{T_o \ (F_n - 1)}{G_1 G_2 \ldots G_{n-1}}$$

Equation 6-18

$$T_e = T_{e1} + \frac{T_{e2}}{G_1} + \frac{T_{e3}}{G_1 G_2} + \ldots + \frac{T_{en}}{G_1 G_2 \ldots G_{n-1}}$$

where
T_{e1} = Noise equivalent temperature of first stage,
T_{e2} = Noise equivalent temperature of second stage,
T_{en} = Noise equivalent temperature of n stage.

NOISE EQUIVALENT
TEMPERATURE OF TWO STAGES IN PARALLEL

Earlier we stated Equation 6-12 which for two parallel stages was,

$$F = 1 + \frac{N_{n1} + N_{n2}}{N_i (G_1 + G_2)}$$

We know the following to be true,

$$\frac{N_{n1}}{N_i G_1} = F_1 - 1 \qquad N_{n1} = N_i G_1 (F_1 - 1)$$

$$\frac{N_{n2}}{N_1 G_2} = F_2 - 1 \qquad N_{n2} = N_i G_2 (F_2 - 1)$$

By substitution,

$$F = 1 + \frac{N_i G_1 (F_1 - 1) + N_i G_2 (F_2 - 1)}{N_i (G_1 + G_2)}$$

$$= 1 + \frac{G_1 (F_1 - 1) + G_2 (F_2 - 1)}{(G_1 + G_2)}$$

$$= 1 + \frac{G_1 F_1 - G_1 + G_2 F_2 - G_2}{G_1 + G_2}$$

$$= 1 + \frac{-G_1 - G_2 + G_1 F_1 + G_2 F_2}{G_1 + G_2}$$

Equation 6-19

$$= \frac{G_1 F_1 + G_2 F_2}{G_1 + G_2}$$

The noise figure may be written in terms of noise equivalent temperature by remembering:

$$F = 1 + T_e/T_o$$

By equating this to Equation 6-19, we obtain:

$$F = 1 + \frac{T_e}{T_o} = \frac{G_1 F_1 + G_2 F_2}{G_1 + G_2}$$

$$\frac{T_e}{T_o} = \frac{G_1 F_1 + G_2 F_2}{G_1 + G_2} - 1$$

Equation 6-20

$$T_e = T_o \left[\frac{(G_1 F_1 + G_2 F_2)}{G_1 + G_2} - 1 \right]$$

Example 6-19

Three amplifier stages have the following data. Compute the overall T_e. Assume Cascade Systems.

	Amplifier #1	Amplifier #2	Amplifier #3
Power Gain	10	10	4
Noise Equi. Temp.	580	810	500

$$T_e = T_{e1} + \frac{T_{e2}}{G_1} + \frac{T_{e3}}{G_1 G_2}$$

$$T_e = 580 + \frac{810}{10} + \frac{500}{10(10)}$$

$$T_e = 580 + 81 + 5$$

$$T_e = 666 \text{ degrees Kelvin.}$$

Example 6-20

Two amplifier stages have the following data. Compute the overall noise equivalent temperature. The reference temperature is 300 degrees Kelvin. Assume parallel connection.

	Amplifier #1	**Amplifier #2**
Power Gain	10	7
Noise Figure	5	10

$$T_e = T_0 \left[\frac{(G_1 F_1 + G_2 F_2)}{G_1 + G_2} - 1 \right]$$

$$= 300 \left[\frac{((10)\,(5) + (7)\,(10))}{10 + 7} - 1 \right]$$

$$= 300 \left[\frac{(50 + 70)}{17} - 1 \right]$$

$$= 300[(120/17) - 1]$$

$$\cong 1818 \text{ °K}$$

SUMMARY

1. As a signal passes through an electrical system the signal is multiplied by the system gain. However, when noise passes through the system, the output noise is not just the input noise multiplied by the gain; instead, it is the input noise multiplied by the gain plus noise from the system itself.

2. The signal-to-noise ratio at the input of a system is degraded as it passes through a system.

At the output in an actual system, the signal-to-noise ratio will be smaller than at the input. A measure of the degrading of signal-to-noise ratio from the input to the output of a system is called noise figure. Expressed as a formula we have:

$$F = \frac{\dfrac{S_i}{N_i}}{\dfrac{S_o}{N_o}}$$

3. When stages are connected in cascade, the first stage usually has a large part in determining the noise figure of the overall system.

4. For very small noise figures it is sometimes difficult to conceive what is actually meant. It is sometimes more understandable to talk in terms of a noise equivalent temperature. Noise equivalent temperature may be stated by the formula:

$$T_e = T_o (F - 1).$$

5. Noise figure has a lower limit of unity or 0 dB and an upper limit of infinity or infinite dB.

6. Noise equivalent temperatures has a lower limit of zero degrees Kelvin while its upper limit is infinity degrees Kelvin.

7. When evaluating the noise figure of an electrical circuit mathematically, we calculate the signal-to-noise ratio at the input and output of the system and place them in the formula given in summary statement two.

PROBLEMS AND QUESTIONS

1. Compute the noise figure of a circuit if the noise output is 70 μW and input noise is 15 μW. The power gain is 3.

2. If network noise is 40 μW, power gain is 4 and input noise is 2 μW, what is the noise figure?

3. For problem two, what is the minimum output noise that could occur?

4. Find the noise figure in dB if F = 1.2, 10, and 20.

5. A network has a network noise of 15 μW, a power gain of 3 and an input noise of 4 μW. What is the output noise and the noise figure?

6. You are given two networks in cascade. Network one has

a noise figure of 6 and a power gain of 10. Network two has a noise figure of 10 and a power gain of 3. What is the overall noise figure?

7. Given the following data, compute the overall noise figure. Use the cascade situation.

Network #	1	2	3	4
Noise Figure	10	15	5	20
Power Gain	3	6	10	12

8. The output of a system composed of two identical stages in cascade is 50 μW. Compute the noise figure of each stage if the power gain of the entire system is 12 and F is 3.

9. Two networks are connected in parallel. Network one has a noise figure of 10 with a power gain of 7. Network two has a noise figure of 10 and a power gain of 6. Compute the overall noise figure.

10. Compute the noise figure of a system if the noise equivalent temperature is 29 degrees Kelvin and the reference temperature is 290 degrees Kelvin.

11. A network is made in three stages with the following data (assume in cascade):

Network	1	2	3
N.E. Temp.	678 °K	325 °K	470 °K
Gain	10	20	15

Find the overall noise temperature and noise figure.

12. Two networks of noise equivalent temp. equal to 150 degrees Kelvin and 300 degrees Kelvin respectively, have a total gain of 17 when placed in cascade. Compute the overall noise figure if the gain of each network is the same.

13. Repeat problem 12 for networks in parallel.

Chapter 7

Noise Measuring
Instruments and Equipment

F OR THE PAST SIX CHAPTERS WE HAVE BEEN DISCUSSING noise. Now let us look at some of the instruments and pieces of equipment that are used in noise measurements. In chapter eight we will learn how various noise measurements are performed.

VOLTMETERS

To measure noise voltages, we will need very sensitive voltmeters since we may be dealing with voltages in the range of fractions of microvolts.

The regular voltmeter found in the laboratory is usually not capable of making noise measurements due to its limited sensitivity. Also, these meters are normally built for measuring the rms value of sine waves, and since noise has an rms level much different than a sine wave, errors may result.

It is possible however to make noise measurements with one of these meters if we know the proper correction factor to use in a noise measurement.

In the average reading voltmeter, the waveform to be measured is first rectified by either a full or half wave rectifier and applied to a dc meter. The dc meter responds to the average level present in the rectified signal. The scale of the dc meter is usually calibrated in regards to the rms value of a sine wave. For example, suppose a sine wave of 10 volts peak was applied to an average reading voltmeter as shown in Fig. 7-1. If the meter had a full wave rectifier present, then the dc meter would be affected by an average level of approximately 6.36 volts. (The average level of a full wave rectified sine wave is the peak value multiplied by $2/\pi$.) The scale of the meter would be calibrated to read the rms level for a 10-volt sine wave or 7.07 volts. This means that the meter scale reads 1.11 times the average level of the incoming sine wave. In Fig. 7-2 we see a schematic of a basic average responding voltmeter. A full wave bridge rectifier has been used in construction of the circuit. When the input voltage swings positive, diodes D_1 and D_4 conduct, allowing current to flow through the current meter. As the input swings negative, diodes D_2 and D_3 can conduct and current flows through the current meter. Over one cycle of operation, the current which flows through the dc meter is related to the average voltage level in the rectified waveform. Also, note that the circuit is built so that the current always flows in the same direction—through the meter. Resistor R limits the current through the meter and fixes the highest current for the meter, when the highest allowable input voltage occurs. This is because the dc current meter has a full scale deflection current which must not be exceeded or meter damage may occur. The meter is usually much more complex in construction than this, since the meter may be called on to measure extremely small voltages.

Suppose a 10-volt square wave had been applied to the average responding voltmeter, the full wave rectified average of the square wave would be the same as its peak value. This means that the current which would flow in the current meter would be related to the 10-volt peak of the waveform. See Fig. 7-3 for a visual description. The meter would then respond giving a reading for the rms level of 11.1 volts since the meter scale reads 1.11 times the average level. However, for a square wave the rms level is the same as the peak value. In this situation, the meter reads high.

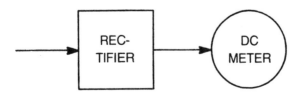

Fig. 7-1. Basic average reading voltmeter diagram.

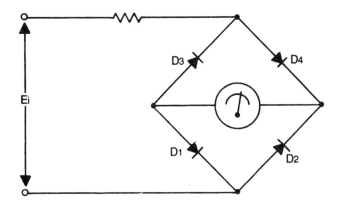

Fig. 7-2. Schematic of an average responding voltmeter.

Therefore, an average responding voltmeter will give an rms level which is in error for any waveform, other than a sine wave. To correct for this discrepancy, we must multiply the reading by a correction factor.

When measuring a square wave's rms level by a full-wave average responding voltmeter, the scale reading must be multiplied by 0.9. In the example illustrated earlier, the meter read 11.1 volts. If we multiplied the meter reading by 0.9 we would have obtained 11.1 × 0.9 = 9.99 volts, which is approximately the rms level of a 10-volt peak square. In general, we may state a formula for the corrected value by Equation 7-1.

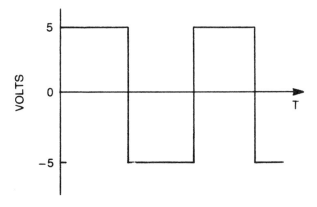

Fig. 7-3. Waveform for Example 7-1.

Equation 7-1

$$X_{CR} = (X_M)(C.F)$$

where

X_{CR} is the corrected value,
X_M is the meter value,
C.F is the correction factor.

We must note that the correction factor depends on the type of meter being used and the type of waveform being measured.

Example 7-1

The waveform shown in Fig. 7-3 is applied to a half-wave average responding meter. Compute the correction factor required for proper measurement of the rms level.

A meter which responds to the half-wave average of the incoming waveform and is calibrated for a sine wave will have scale markings based on the fact that the rms level of a sine wave is .707 of the peak value and the half-wave average is 0.318 of the peak. Therefore, if a 10-volt sine wave is applied, the meter responds to the average level of 3.18 volts. But the scale reading indicated is 7.07 volts. (The scale will read 7.07/3.18 = 2.22 times higher.)

For the five-volt square wave shown, the half-wave average is 2.5 volts. The meter will read 2.5 × 7.07/3.18 or 5.55 volts. The correction factor to use is the value which when multiplied by 5.55 will give the correct reading of five volts. The correction factor is then C.F. = X_{CR}/X_M = 5/5.55 = 00.9

If noise would have been placed on the meter input, the meter would not read the rms level of the noise. It is very difficult to state a correction factor for the noise since there are so many different possible forms for noise. As a rule of thumb, to calculate the rms noise voltage from a full-wave average responding meter, multiply the meter reading by 1.15. This correction factor is rather difficult to prove by formulas and is stated from the results of various experiments which have been performed.

Example 7-2

Noise is impressed on an average responding full wave voltmeter. The meter reads 40 μV. What is the rms value of the noise waveform?

$$V_n = V_m \ (1.15)$$

$$= 40 \ (1.15)$$

$$= 46 \ \mu V.$$

TRUE RMS VOLTMETER

A true rms voltmeter reads the actual rms value of the waveform applied regardless if it is a sine wave, square wave, sawtooth, exponential or any other type of waveform. This also includes noise voltages. A 10-volt peak sine wave when applied to a true rms voltmeter will give a reading of 7.07 volts. Likewise, a 10-volt peak square wave will give a reading of 10 volts on the meter.

True rms meters have been constructed using circuits containing a thermocouple. The voltage to be measured produces a cur-

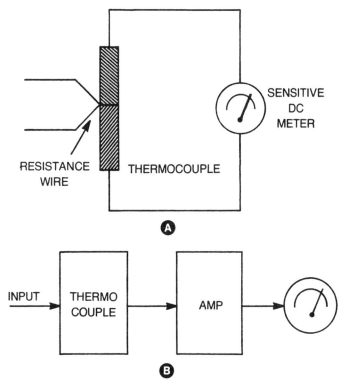

Fig. 7-4(A). Basic true rms meter using a thermocouple. (B) Thermocouple true rms meter with amplifier.

rent in a short piece of resistance wire. The wire heats and is connected to a thermocouple which consists of a junction of two different metallic wires that generate a voltage when heated. The output of the thermocouple is sent to a sensitive dc meter which gives an indication of the alternating current passing through the meter. The amount of heat generated by the passage of current through the meter is directly proportional to the square of the effective (rms) current of the applied voltage to the meter input terminals. A basic meter which can be used to measure the true rms level of a voltage is shown in Fig. 7-4A. For noise measurements, an amplifier must precede the basic meter. This is illustrated in Fig. 7-4B.

COMPARISON BETWEEN AVERAGE RESPONDING AND TRUE RMS VOLTMETERS

True rms voltmeters are more accurate and expensive than a comparable average reading voltmeter and do not require correction factors. When noise measurements are made to determine a ratio, an average responding meter may be used since the following relation holds true as Equation 7-2.

Equation 7-2

$$\frac{V_2}{V_1} = \frac{C.F.(V2')}{C.F.(V1')}$$

where V_1 and V_2 are the correct values and V_1' and V_2' are the indicated meter values. However, when direct noise measurements are required a correction factor must be used in the measurement procedure.

Average responding meters should never be used when a complex signal such as a sine wave and a noise waveform combination are to be measured. A true rms meter should be used in this situation.

NOISE GENERATORS

A noise generator is a device which delivers a waveform whose instantaneous amplitude is determined at random. Random noise measurements are used in many scientific areas. For example, random noise resembles the vibrations that mechanical systems are confronted with. Therefore it is used frequently in vibration testing.

Various accoustical measurements are made simpler as random noise can abe used to smooth response curves which are difficult to describe. In electrical systems random noise is used for such tests as noise figure and crosstalk measurements.

Figure 7-5 shows a basic noise generator. A zener diode is used as the noise source. When operated near the knee of the zener characteristic, the zener diode produces a fairly large amount of noise.

Typical noise voltages generated by zeners yield an rms noise level of 1.5 millivolts for every 50 ohms of dynamic zener resistance. Zeners may generate noise with frequency components in the range from 50 kHz to 250 kHz. The zener diode has its noise voltage amplified by the transistor to produce a strong noise signal at the output. Such a circuit is examined in Experiment 2. It is noted that noise can be observed on an oscilloscope by using the simple circuit shown in Fig. 7-6 for many types of zener diodes. However, some diodes may require amplification of their noise voltage before it is applied to an oscilloscope.

A more sophisticated noise generator shown in Fig. 7-7 also uses a zener diode. Noise generated in a zener diode is amplified and applied to a mixer. In the mixer, all of the frequencies in the

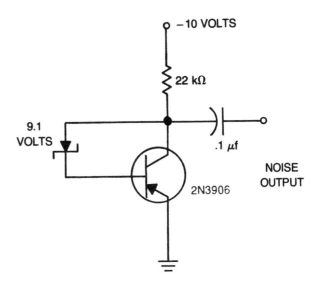

Fig. 7-5. Basic noise generator using a zener diode.

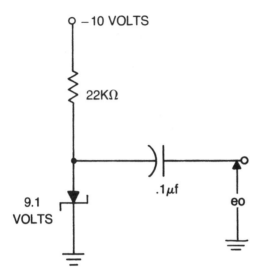

Fig. 7-6. Generation of noise using a zener diode.

zener diode noise are beat with a sine wave from a local oscillator. The reason for doing this is that the zener generates noise from 50 kHz to 250 kHz. We would like noise components extending down to near dc, so we must convert the zener noise frequency spectrum by using a local oscillator. The process is essentially the same as what occurs in a superhetrodyne receiver when the incoming

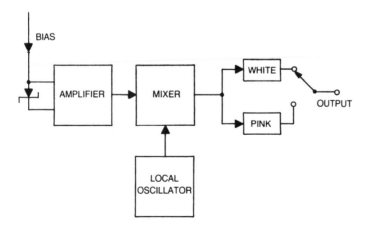

Fig. 7-7. Typical noise generator with its associated sections.

rf signal is hetrodyned down to the i-f signal. The output of the mixer becomes noise of a different frequency makeup. For example, the new frequency spectrum may be from near dc to 150 kHz. This noise is sent into a series of filters. One filter produces an output which is regarded as white noise since the frequency makeup has a flat spectrum. Another filter allows an output which has a "pink frequency spectrum." This is because the amplitudes of the lower frequency spectrum components are stronger than the higher frequency components.

More detailed noise generators may include other types of filters. One filter that is quite popular produces an output spectrum that resembles the frequency distribution of speech. This type of filter gives a rising response at low frequencies until it peaks somewhere in the vicinity of 500 to 800 Hz. Then the response falls and reaches a relative response of – 25 dB at 5 kHz. This is very similar to the human voice spectrum and can be used for testing many types of audio circuits and components. Figure 7-8 shows the appearance of a filter characteristic that can perform this function.

After the filter section, an amplifier may be incorporated which increases the noise level to a more convenient value. It also serves to buffer the generator from various loads.

When a temperature limited vacuum tube diode has current flowing through it, a shot noise current will appear superimposed on the plate current. This current can be easily predicted using the equation for shot noise current. A dc meter is present in the circuit

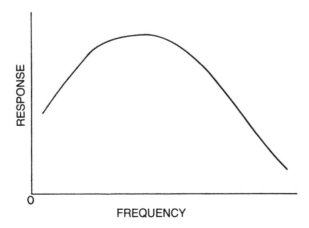

Fig. 7-8. Filter to produce a frequency spectrum similar to that of speech.

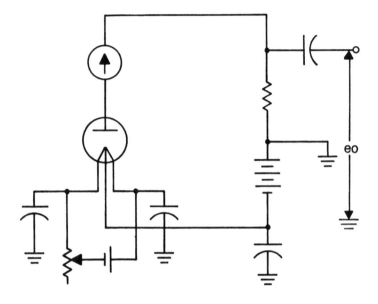

Fig. 7-9. Noise generator using vacuum tube.

to measure plate current. The filament is heated by current from a small battery placed in series with a rehostat and the filament. As the rehostat is varied the amount of filament current can be varied and the amount of plate current can be changed. The schematic of this generator is shown in Fig. 7-9. In chapter eight there will be a discussion of how this generator is used to perform noise measurements. Although all vacuum tube diodes produce shot noise current when biased properly, there are vacuum tube diodes available that have been manufactured specifically for use in noise generators. The noise generated in these types of generators may have frequency components in the Megahertz region.

SPECTRUM ANALYZER

A spectrum analyzer produces a graphical plot of amplitude versus frequency for a signal. Although the physical appearance may be similar to an oscilloscope, the oscilloscope produces a graphical plot of signal voltage versus time.

Use of A Spectrum Analyzer

The spectrum analyzer can be used to make measurements

such as modulation, signal and system bandwidth, spurious signal generation and noise spectrum analysis.

Applications for spectrum analysis exist in many different technical areas including instrumentation, navigation and various forms of communications such as AM, FM and pulse modulation.

Spectrum Analyzer Operation

In Fig. 7-10 is shown a diagram of a spectrum analyzer. A spectrum analyzer block diagram looks very similar to a block diagram of a superhetrodyne receiver. The major differences are the sawtooth generator is connected to the local oscillator, and there is a cathode ray tube.

The input signal to be investigated in regards to its frequency spectrum is applied to the mixer input circuit. The horizontal position on the screen is controlled by the size of the sawtooth voltage. Likewise, a voltage-tuned local oscillator varies its frequency as the sawtooth voltage varies. Refer to Fig. 7-11 for a visual description of the operation of the voltage tuned oscillator. As the sawtooth voltage increases, the local oscillator frequency increases. The sawtooth voltage also serves the same purpose as the horizontal sweep voltage in an oscilloscope. Since the sawtooth voltage size controls the local oscillator signal frequency, we can say that the local oscillator is swept.

The incoming signal, which is applied to the mixer input, has signals present in its frequency spectrum beat by the swept local oscillator output and given an intermediate signal frequency. This

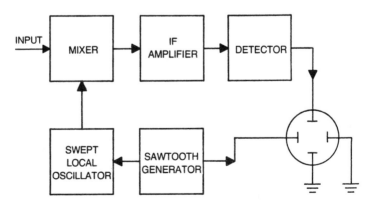

Fig. 7-10. Spectrum analyzer block diagram.

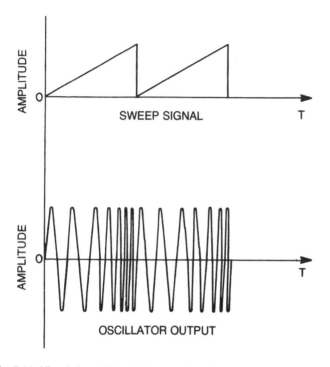

Fig. 7-11. Visual description of the operation of a voltage tuned oscillator.

signal only exists when a frequency component is present in the input signal. The intermediate signal is amplified and detected and applied to the vertical plates of the cathode ray tube (CRT). Since the size of the sawtooth voltage controls the horizontal position of the electron beam on the CRT screen and also the local oscillator frequency, whenever an intermediate signal is produced it will result in a vertical deflection at a particular horizontal position. This horizontal position may be calibrated in terms of frequency. The result is that we observe a plot of amplitude versus frequency for the frequency spectrum on the screen. Figure 7-12 illustrates the frequency and time spectrum which would be given by a spectrum analyzer and oscilloscope respectively for various types of signals. In Fig. 7-12A we see an AM waveform with single tone sinusoidal modulation in both the frequency and time domains. Note the carrier and two side frequency components. Likewise in Fig. 7-12B a typical FM process is indicated for single tone modulation. Figure 7-12C indicates the frequency and time domains for a square wave. Notice how the components in the frequency domain decrease

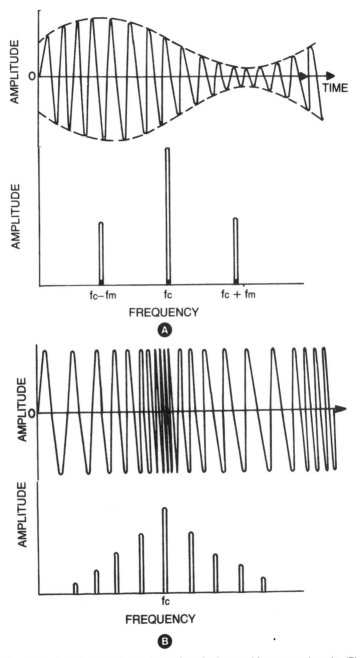

Fig. 12(A). Amplitude modulated waveform in time and frequency domain. (B) Frequency modulation waveform in time and frequency domains.

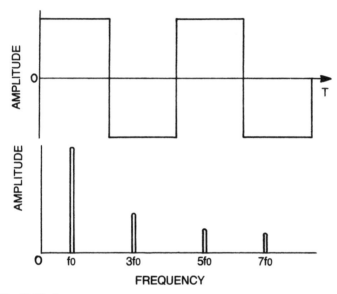

Fig. 12 (C). Square wave in the time and frequency domains. (Continued from page 187.)

in amplitude for higher frequencies. Also, components exist only at odd multiples of the fundamental.

Spectrum analyzers come in various sizes and many different forms. Some may measure frequency over a small range while others may make frequency measurements over a large range. Screens may be a few inches in width or as large as the screen on a sizeable television. Some spectrum analyzers are capable of showing a range of frequencies on the screen as much as several gigahertz. Some spectrum analyzers appear in the form of plug-ins where they can replace the vertical amplifier of a regular oscilloscope. When finished with the analyzer, the vertical amplifier is simply replaced into the oscilloscope. Other spectrum analyzers are complete units which may cost thousands of dollars.

Spectrum Analyzer Terms

Let us now look at some of the more important spectrum analyzer terms.

Center frequency range is defined as the range of frequencies which can be displayed at the center of the reference coordinate. The frequency which corresponds to this center of the reference coordinate is called the center frequency.

188

Drift is defined as long term frequency changes caused by changes or instabilities in the spectrum analyzer.

Minimum discernible signal is the input signal level which gives a presentation on the spectrum analyzer screen, where the signal is just identifiable from the noise.

Resolution is the analyzer's ability to display adjacent signals within a frequency span. See Fig. 7-13 for a pictorial representation.

Window refers to the frequency range which can be seen on the screen at one particular setting. If the window is measured in centimeters and the horizontal setting is given in kilohertz per centimeter, the window width is given by Equation 7-3.

Equation 7-3

$$W = (S)\ (L)$$

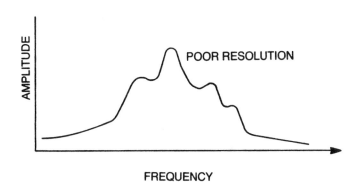

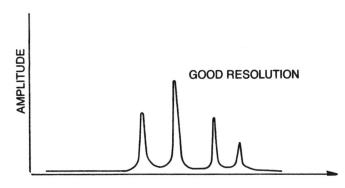

Fig. 7-13. Pictorial representation of resolution.

where

W is the window in kilohertz,

S is the analyzer setting in kilohertz per centimeter,

L is the window width.

Example 7-3

A spectrum analyzer is set for a horizontal setting of 5 kHz per cm. If the screen width is 10 cm, compute the window width.

$$W = (S)(L)$$
$$= (5 \text{ kHz})(10 \text{ cm})$$
$$= 50 \text{ kHz}.$$

Dispersion is a measure of how broad a spectral line on the analyzer happens to be. Well designed analyzers have narrow spectral lines or a small amount of dispersion. The wider or larger a spectral line width, the more difficult it becomes to distinguish signals from each other.

Logarithmic Gain allows small signals such as noise to be given more gain than large signals for presentation on the analyzer screen. The result is that the small signals can be carefully studied in respect to the large signals.

These are just a few of the terms used to describe spectrum analyzers. If you have further interest, it is advised to consult a manufacturers equipment catalog.

SUMMARY

1. When making accurate noise measurements, it is always best to use a true rms meter.

2. Average responding meters can be used in noise measurements provided a correction factor is employed.

3. Noise generators may be used to produce white, pink, and various other types of noise processes by incorporating a filter into the output circuitry of the generator.

4. A very popular noise generator uses a vacuum tube diode to generate shot noise. We have learned in earlier chapters that shot noise may be easily predicted by basic formulas.

5. An excellent source of noise for a noise generator is a zener diode.

6. A spectrum analyzer produces a plot of amplitude versus frequency. The amplitude may be indicative of voltage or relative power.

PROBLEMS AND QUESTIONS

1. Explain in your own words the difference between an average responding and a true rms voltmeter.

2. A square wave of peak value equal to 20 volts is applied to a full-wave average responding voltmeter. Compute the correction factor required to convert the rms value indicated on the meter to the true rms level.

3. The same waveform in problem two is applied to a true rms voltmeter. What is the reading indicated on the voltmeter scale?

4. A noise process having a value of 0.78×10^{-6} volts squared is applied to the following instruments. Give the voltages indicated on the meters.

      a. average responding voltmeter (full wave)

      b. true rms voltmeter

5. Explain the operation of a spectrum analyzer. Use diagrams to help in your explanation.

6. Draw a block diagram of a noise generator which uses a frequency mixer. Explain the purpose that the mixer serves.

7. Research question: What are the differences between a spectrum analyzer and an oscilloscope? What are the differences between a spectrum analyzer and a superhetrodyne receiver?

Chapter 8

Noise Measurements

I N THE PREVIOUS CHAPTER WE LEARNED ABOUT VARIOUS IN-
struments and pieces of equipment used in making noise measurements. Now let us address the subject of how we make the various noise measurements.

MEASUREMENT OF NOISE EQUIVALENT RESISTANCE

Noise equivalent resistance is measured by first taking the electronic device whose R_{eq} is to be determined and biasing the device
to some particular operating point. Then the input element of the
device is grounded. This may be done with a capacitor to produce
an ac ground in the case of a device that needs dc current in the
input element. An example would be a junction transistor in which
the base would be the input element. Low noise resistors should
be used for the external biasing circuit to prevent obtaining a value
of R_{eq} that is related to external components. The output noise of
the circuit is then measured with a true rms voltmeter or an average reading ac voltmeter. Under most conditions a high gain, low
noise voltage amplifier will be required to precede the voltmeter
since we will be dealing with extremely small voltages. The diagram for the circuit measurement set up is shown in Fig. 8-1. The
noise output is caused by the noise generating property of the active device. We may think of this noise generating property as that

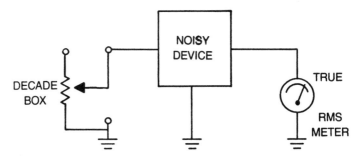

Fig. 8-1. Measurement setup for noise equivalent resistance.

due to R_{eq} as shown in Fig. 8-2. The output noise has a rms voltage squared given by Equation 8-1.

Equation 8-1

$$E_{no1}{}^2 = (4\ kTR_{eq}B)\ A_v{}^2$$

Now the ground on the input element is removed and an adjustable resistance is connected between the input element and ground as shown in Fig. 8-2. The resistor is adjusted until the output noise voltage squared doubles. This is indicated by an increase in the output noise voltage by the square root of two. In other words, $E^2_{no2} = \sqrt{2}E_{no1}$. Under this condition $E^2_{no2} = 2E^2_{no1}$. Since the increase in output noise is due to the adjustable resistance, we may write Equation 8-2.

Equation 8-2

$$E^2_{no2} = 2\ E^2_{no1} = [4kT\ (R_{eq} + R_a)B]\ A_v{}^2$$

Req REPRESENTS NOISE GENERATING
PROPERTY OF DEVICE

```
        Req ⟩          DEVICE          NOISE
                    THOUGHT OF         OUTPUT
                        AS
                    NOISELESS
```

Fig. 8-2. The noise associated with the device may be thought of as coming from a resistor R_{eq}.

R_a is the adjustable resistor value.

We may also divide Equation 8-2 by Equation 8-1 to obtain Equation 8-3.

$$2 = \frac{E^2_{no2}}{E^2_{no1}} = \frac{(4kT\,(R_{eq} + R_a)B)A_v^2}{(4kT\,R_{eq}B)\,A_v^2}$$

Equation 8-3

$$2 = \frac{R_{eq} + R_a}{R_{eq}}$$

Solving Equation 8-3 for R_a we have

$$R_a = R_{eq}.$$

Therefore, for an increase in output noise voltage by the square root of two, we can read the adjustable resistor directly for the equivalent noise resistance.

Should we desire not to use an increase by the square root of two, we can use a more general formula, Equation 8-4.

Equation 8-4

$$\frac{E_{no2}^2}{E_{no1}^2} = 1 + \frac{R_a}{R_{eq}}$$

We may solve this equation for R_{eq} which Equation 8-5 states.

Equation 8-5

$$R_{eq} = R_a\left(\frac{E_{no1}^2}{E_{no2}^2 - E_{no1}^2}\right)$$

Example 8-1

An active device is biased and its input element is grounded. The output noise voltage is 50 μV. An adjustable resistor is then connected between the input element and ground. The resistor is adjusted until the output noise voltage has doubled. The adjustable resistor reads 10 kΩ. What is the equivalent noise resistance of the active device?

1. $E_{no1} = 50 \mu V$

2. $E_{no2} = 2 E_{no1} = 2 (50 \mu V) = 100 \mu V.$

3. $R_{eq} = R_a \dfrac{E_{no1}^2}{E_{no2}^2 - E_{no1}^2}$

$$= (10 \text{ k}\Omega) \left(\frac{(50 \mu V)^2}{(100 \mu V)^2 - (50 \mu V)^2} \right)$$

$$= 10 \text{ k}\Omega \left(\frac{(2500 (\mu V)^2)}{(10{,}000 - 2500) (\mu V)^2} \right)$$

$$= 10 \text{ k}\Omega \frac{(2500)}{7500} = 3.33 \text{ k}\Omega.$$

MEASUREMENT OF NOISE EQUIVALENT BANDWIDTH

We have already learned how to calculate noise equivalent bandwidth. However, how do we measure such a quantity? Since noise equivalent bandwidth is related to frequency response, it only seems right that this response be included in the measurement procedure. We will now examine two methods to measure this bandwidth.

Signal Generator Method

In Fig. 8-3 we see the basic setup for this measurement procedure. A signal generator is connected to the system in question and ac voltmeters are connected across the input and output terminals of the system. The system has the appropriate biasing potentials connected to it, and then a frequency response test is made on the system. The generator is set to some frequency f_1 and the input voltage and output voltage are measured. The generator is set to some new frequency f_2 and the input voltage is adjusted to maintain the same value as for frequency f_1. The output voltage is again measured and recorded. This procedure is repeated until enough information has been gathered to plot accurately a voltage gain squared versus frequency graph. The graph is plotted on lin-

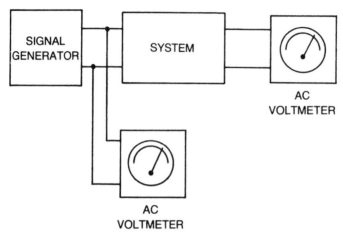

Fig. 8-3. Setup for the measurement of noise equivalent bandwidth.

ear graph paper. No other type of graph paper may be used. A typical plot is shown in Fig. 8-4A. The procedure then involves determining the area under the voltage gain squared versus frequency curve. To do this we then divide the area up into rectangles and triangles as shown in Fig. 8-4B. We will then calculate the area of each of these rectangles and triangles and sum them together.

We then determine the voltage gain at mid frequencies if the response is bandpass or at low frequencies if the response is low pass. We will call this value A_p. The noise equivalent bandwidth is then determined by using Equation 8-6.

Equation 8-6

$$R_{eq} = \frac{\text{Summation of individual areas under square of voltage gain curve}}{(A_p)^2}$$

$$= \frac{\displaystyle\sum_{i=1}^{n} A_i}{(A_p)^2}$$

where A_i represents the individual areas.

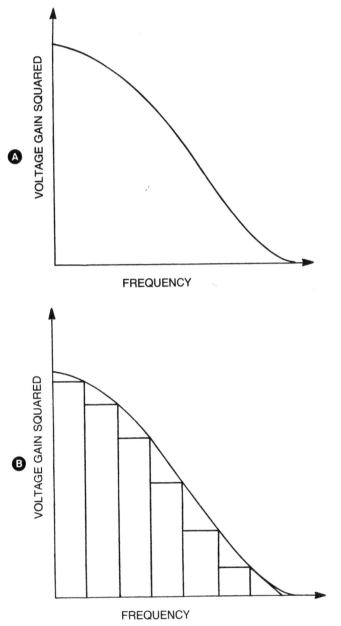

Fig. 8-4(A). The graph shows voltage gain squared versus frequency. (B) Dividing area into rectangles and triangles.

Spectrum Analyzer Method

Another method of measuring noise equivalent bandwidth is to use a wide band spectrum analyzer and an impulse generator. An impulse generator gives an output of extremely narrow pulses. These pulses produce a relatively flat frequency spectrum. This spectrum is presented to the input of the system and the output frequency spectrum becomes altered by the frequency response of the system. The spectrum analyzer is connected to the output to observe this frequency response. A photograph is then taken, and the curve is geographically squared as in Fig. 8-5. The area is then determined as in the signal generator method and the bandwidth is computed using Equation 8-6.

We note that a white noise generator could have been used instead of an impulse generator. White noise also has a flat frequency spectrum.

An experiment is given at the end of this book to determine noise equivalent bandwidth. An example in determining this bandwidth is given in the discussion for this experiment.

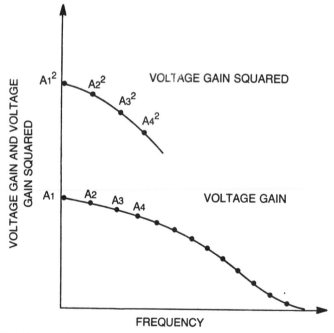

Fig. 8-5. Graphically squaring the voltage gain curve.

SIGNAL-TO-NOISE RATIO

When making signal-to-noise ratio measurements, we must perform measurements of signal and noise. The main problem exists since when we are measuring signal, we also are measuring a certain amount of noise, too. In many instances, we need not worry about this especially when the signal-to-noise ratio is expected to be larger than 10.

The procedure for measurement is as follows:

1. Connect the circuit as shown in Fig. 8-6. Apply power to the circuit and connect the signal generator to the system. Adjust the signal generator to the desired input voltage value, and measure the signal voltage at the test point with a narrow band voltmeter. Call this voltage e_s.

2. Turn off the signal generator but leave it connected to the circuit. With a true rms voltmeter, measure the amount of noise voltage at the test point. Call this voltage e_n.

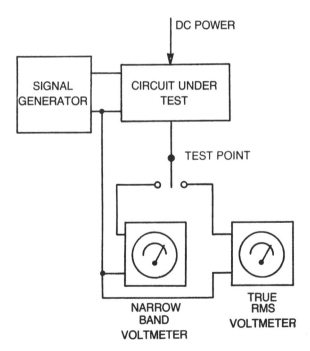

Fig. 8-6. Setup for measuring signal-to-noise ratio.

3. Calculate the signal-to-noise power ratio as follows:

$$\frac{S}{N} = \frac{e_s^2}{e_n^2}$$

Why use a narrow band voltmeter to measure the signal? Assuming that the signal being used is a sine wave, the narrowband voltmeter will measure accurately the signal strength while placing a small bandwidth on the noise at the test point. Even though we can't separate the noise and signal from each other, we can reduce the effect of noise being added on to the signal by using the narrow band voltmeter.

Suppose we measure the signal and noise voltage with the same voltmeter. We will then be dealing with a situation in which we must consider the effect of a signal and noise component to noise ratio. For example, with the signal applied, the voltage measured at the test point will be the square root of the sum of the squares of the signal and noise voltage "e." We may then write this voltage as Equation 8-7.

Equation 8-7

$$e^2 = e_n^2 + e_s^2$$

The signal generator is then turned off and we measure the noise at the test point; e_n. The ratio now becomes that of Equation 8-8.

Equation 8-8

$$\frac{S + N}{N} = \frac{(e_n^2 + e_s^2)}{e_n^2} = 1 + \frac{e_s^2}{e_n^2}$$

Example 8-2

The signal and noise combined give a noise signal of 30 μV. The noise voltage measured is 10 μV. What is the signal-to-noise ratio?

$$\frac{S + N}{N} = \frac{(30 \ \mu V)^2}{(10 \ \mu V)^2} = \frac{900}{100} = 9$$

$$\frac{S}{N} = \frac{S + N}{N} - 1 = 9 - 1 = 8.$$

MEASUREMENT OF NOISE EQUIVALENT VOLTAGE

This measurement is fairly simple in that it requires a voltage gain measurement and a noise voltage measurement. The procedure is to properly bias the electronic device or system to the operating point desired. Then the voltage gain of the network is measured. The output noise voltage is determined with a true rms meter when the input is grounded. The voltage gain is measured by inserting a signal from a signal generator into the input of the active device and measuring the input voltage e and output voltage e_o. The voltage gain A_v is then calculated using Equation 8-9.

Equation 8-9

$$A_v = e_o/e_i$$

The output noise voltage is then divided by the voltage gain. This reflects the noise voltage to the input circuit. Equation 8-10 gives the formula for noise equivalent voltage.

Equation 8-10

$$E_{neq} = V_{no}/A_v$$

Refer to Fig. 8-7 for this measurement procedure.

Example 8-3

The output noise voltage of a device, when properly biased and

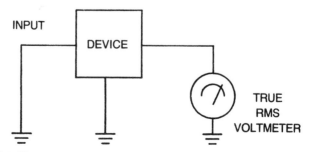

Fig. 8-7. Measurement of noise equivalent voltage.

input circuit grounded, is 40 μV. The voltage gain is 16. Compute the noise equivalent voltage.

$$E_{neq} = \frac{V_{no}}{A_V} = \frac{40\ \mu V}{16} = 2.5\ \mu V.$$

MEASUREMENT OF NOISE FIGURE

Noise figure may be measured with a signal generator, two regular ac voltmeters and a true rms voltmeter. The equipment setup is shown in Fig. 8-8. The procedure is as follows:

1. Measure the noise equivalent bandwidth of the system as given earlier in this chapter by making a response curve of the square of voltage gain versus frequency.
2. Determine the midband voltage gain of the amplifier and square that value.
3. Calculate the thermal noise voltage due to the generator's internal resistance using the formula
$e_n^2 = 4kTR_sB_{eq}$.
4. Measure the output noise voltage of the amplifier using a true rms voltmeter. Square this value and call it e_{no}^2.
5. Calculate the noise figure using Equation 8-11.

Equation 8-11

$$F = \frac{e_{no}^2}{(4kTR_sB_{eq})\,(A_v^2)}$$

Signal Generator Method #2

An alternate method of measuring noise figure is shown by Fig. 8-9. The procedure is as follows:

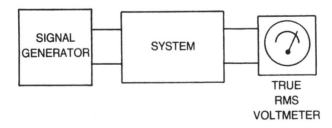

Fig. 8-8. Equipment setup for measurement of noise figure.

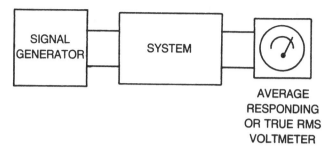

Fig. 8-9. An alternate method of measuring noise figure.

1. Measure the output noise voltage with a true rms voltmeter or an average reading ac voltmeter, when the generator is connected to the input of the system. Make sure that the generator is set for zero output signal voltage. The noise output voltage squared is then given by Equation 8-12.

Equation 8-12

$$e^2_{no} = (4kTR_sB_{eq})(A_v^2) + e^2_{nn}$$

where e^2_{nn} is the extra noise produced by the system.

2. The signal generator is adjusted until the voltmeter reading for the output noise voltage increases by the square root of two. The output voltage squared is then given by Equation 8-13.

Equation 8-13

$$2\,e^2_{no} = (4kTR_sB_{eq} + e^2_{so})(A_v^2) + e^2_n$$

Because the signal generator input has caused the output power to double, we can say that the amount of signal voltage squared present at the output is the same as the output noise voltage squared. This makes the output signal to noise ratio equal to unity. Since noise figure is equal to input signal-to-noise ratio divided by output signal-to-noise ratio, and output signal to noise is unity in this case, then the noise figure is the input signal-to-noise ratio. The input signal to noise is now equal to e^2_{so} divided by $4kTR_sB_{eq}$ where e_{so} is the open circuit voltage of the generator.

3. Disconnect the signal generator from the system and read the open circuit output voltage e_{so}.

4. Calculate the noise figure using Equation 8-14.

Equation 8-14

$$F = e_{so}^2/4KTR_sB_{eq}$$

It is obvious from this formula that we are dealing with relatively small voltages when the noise figure is small. The noise figure is dependent on a thermal noise voltage which normally will be in the fraction of microvolts.

Noise Generator Method

In Chapter 7 we learned about the construction of a noise generator using a vacuum tube diode. Such a generator is an excellent source of white noise, usually up to about 5 MHz. Refer to Fig. 8-10 for the basic setup for equipment. The measurement procedure is as follows:

1. The noise generator resistance R_g is made equal in value to the input resistance R_{in} of the network. Under this condition the available noise power P_n from the generator is supplied to the input terminals of the network in question. The available noise power P_n is given by Equation 8-15.

Equation 8-15

$$P_n = E_g^2/4 R_g$$

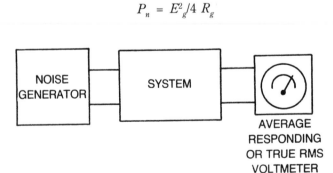

Fig. 8-10. Measuring noise figure using a noise generator.

204

2. The noise generator is adjusted to zero output and the output noise voltage of the network is measured. The output noise power is given by Equation 8-16.

Equation 8-16

$$P_{no} = G\,P_n + P_{network}$$

where P network is the noise component of power produced by the network alone.

3. The noise generator output is adjusted until the output voltage of the network is increased by the square root of two or the power is doubled. The extra power that was added to the input circuit was P. Therefore we may write an equation for the new output power which is Equation 8-19.

Equation 8-19

$$2\,P_{no} = G\,P_n + P_{network} + GP$$

4. We may derive an equation for noise figure, Equation 8-20.

$$F = \frac{N_o}{G\,N_i} = \frac{GP}{G\,N_i} = \frac{N}{N_i}$$

$$= \frac{\dfrac{E_g^{\,2}}{4R_g}}{k\,T\,B_{eq}}$$

Equation 8-20

$$F = \frac{E_g^{\,2}}{4kTR_g\,B_{eq}}$$

It can be shown that E_g is given by multiplying the shot noise current by the resistance R_g. The result is given by Equation 8-21. By substituting Equation 8-21 into Equation 8-20 we obtain Equation 8-22.

Equation 8-21

$$E_g = (2eI_{dc}B)^{1/2} R_g$$

$$F = \frac{(2eI_{dc}B) R_g^2}{4kTR_gB}$$

Equation 8-22

$$F = \frac{eI_{dc}R_g}{2kT}$$

If the temperature is 290 degrees Kelvin then the formula for noise figure becomes that of Equation 8-23.

Equation 8-23

$$F = 20I_{dc}R_g$$

Example 8-4

A noise figure measurement is made with the diode noise generator. The plate current read is 3 mA when the output noise power is doubled, resistor R_g is 50 Ω. The bandwidth of the network is 6 megahertz. What is the noise figure assuming the temperature is 290 degrees Kelvin?

$$F = 20 \, I_{dc}R_g$$
$$F = 20 \, (3 \times 10^{-3}) \, (50)$$
$$F = 3.$$

NOISE EQUIVALENT TEMPERATURE

Noise equivalent temperature may be measured by using a testing system which incorporates thermal noise sources at different temperatures. These sources are resistors. One resistor is maintained at a temperature T_c which we call the cold value. This is accomplished by such methods as circulating a cooling liquid around the resistor. The other resistor is maintained at a temperature T_h which we call the hot value. This is accomplished by keeping the resistor in an oven. Each resistor has the same resistive value. Therefore, the hot resistor will have a larger noise voltage than the cold resistor.

We learned earlier that the noise power available from a resistor is directly proportional to the resistor temperature. The same thing holds true in this measurement procedure, so that the available noise powers presented to the system are dependent on both the hot and cold temperatures.

The noise power at the output of the system is thought of as being produced by an effective noise temperature or equivalent noise temperature at the input. The procedure for measurement is as follows. Also refer to Fig. 8-11.

1. Set the switch to the cold resistor terminal and record the output noise voltage, E_{no1}. Set the switch to the hot resistor terminal and record the new output noise voltage E_{no2}.

2. Compute the power ratio M.

$$M = \left(\frac{E_{no2}}{E_{no1}} \right)^2$$

3. The M factor is also equal to the ratio of the sum of available noise power, due to the hot temperature and equivalent noise temperature, to the sum of available noise power, due to the cold temperature and equivalent noise temperature.

This is stated by Equation 8-24.

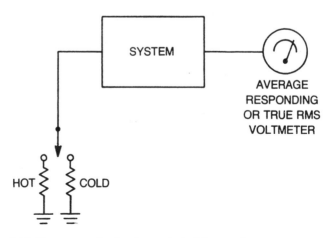

Fig. 8-11. Measurement setup for equivalent temperature.

$$M = \frac{kT_hB + kT_{eq}B}{kT_cB + kT_{eq}B}$$

Equation 8-24

$$M = \frac{T_h + T_{eq}}{T_c + T_{eq}}$$

$$M\,T_c + M\,T_{eq} = T_h + T_{eq}$$

$$M\,(T_{eq}) - T_{eq} = T_h - M\,T_c$$

$$T_{eq}\,(M - 1) = T_h - M\,T_c$$

Solving this Equation for T_{eq}, we obtain Equation 8-25.

Equation 8-25

$$T_{eq} = \frac{T_h - M\,T_c}{M - 1}$$

Example 8-5

Two thermal noise sources are used to measure the noise equivalent temperature. When the cold resistor is connected to the system the output noise voltage is 30 μV. When the hot resistor is connected the output noise voltage is 45 μV. The cold temperature is 17 degrees Centigrade. The hot temperature is 840 degrees Kelvin. Determine the equivalent noise temperature.

1. Convert the cold temperature to degrees Kelvin.

$$T_c = 273 \text{ degrees} + C$$

$$= 273 + 17$$

$$= 290 \text{ degrees Kelvin.}$$

2. Find M.

$$M = \left(\frac{E_{no2}}{E_{no1}}\right)^2 = \left(\frac{45}{30}\right)^2 = (1.5)^2 = 2.25$$

3. Compute T_{eq}.

$$T_{eq} = [(T_h - M\ T_c)]\ (M - 1)$$
$$= [840 - (2.25)\ (290)]/(2.25 - 1)$$
$$= [840 - 652.5]/1.25$$
$$= 187.5/1.25$$
$$= 150\ \text{degrees Kelvin.}$$

Example 8-6

Using Equation 8-25 and the equation for noise figure, derive an equation for noise figure in terms of T_c, T_h and M.

We know that noise figure F is related to noise equivalent temperature by the formula:

$$F = 1 + T_{eq}/T_o$$

where T_o is the reference temperature.

By substituting the formula for T_{eq} into the equation for F we obtain:

$$F = 1 + \cfrac{\dfrac{T_h - M\ T_c}{M - 1}}{T_o}$$

which is also equal to

$$F = 1 + \frac{T_h - M\ T_c}{T_o\ (M - 1)}$$

Example 8-7

If the reference temperature is 290 degrees Kelvin for Example 8-5, what is the noise figure?

$$F = 1 + T_e/T_o$$
$$F = 1 + 150/290$$
$$F = 1.5172$$

The measurements covered in this chapter must be made under conditions that do not introduce large amounts of noise from the outside environment.

SUMMARY

1. In this chapter we have learned how some of the more important noise measurements are made on an electronic system. We must remember that although the block diagrams look fairly simple, the equipment and test circuits may be sophisticated to prevent interference from outside noise sources.

2. Sometimes we may deal with noise measurements which are very difficult to observe on a noise voltmeter. It may be necessary to add additional amplification to read these noise voltages.

QUESTIONS AND PROBLEMS

1. Discuss the measurement procedure for the following:

   a.) noise figure. b.) noise equivalent temperature. c.) noise equivalent bandwidth.

2. What is meant by a spot noise figure? How does average noise figure differ from spot noise figure? (Research Questions)

3. Draw a block diagram for a measurement setup for noise figure.

4. An FET is biased and its gate is grounded. The output noise voltage measured is 20 μV. An adjustable resistor is connected between the gate and ground. The resistor is adjusted until the output noise voltage has tripled. The adjustable resistor reads 5 kΩ. What is the equivalent noise resistance of the active device?

5. A noise equivalent bandwidth test is made on a bandpass filter. The mid-band voltage gain is 20. Compute the noise equivalent bandwidth if the area under the voltage gain squared versus frequency is 50,000 hertz.

6. Compute the equivalent noise voltage of a system that has a voltage gain of 70 and an output noise voltage with input shorted of 50 microvolts.

7. A noise figure measurement is made with a vacuum tube diode noise generator. The input resistance of the system is 300

ohms. Compute the noise figure if the plate current is 4 milliamps when the output noise voltage is increased by a factor of $\sqrt{2}$.

8. Two thermal noise sources of 100 degrees Kelvin and 800 degrees Kelvin are used in a noise equivalent temperature measurement. When the cold resistor is connected to the system the output noise voltage is 40 μV. When the hot resistor is connected the output noise voltage is 80 μV. If the reference temperature is 290 degrees Kelvin, compute the noise equivalent temperature and the noise figure.

Chapter 9

Noise In
Communication Systems

L ET US NOW ADDRESS THE SUBJECT OF NOISE IN THE VARI-
ous types of communication systems as it pertains mainly to
the receiver.

A model of the typical communication system is shown in Fig.
9-1. Noise may be generated in the transmitter, receiver or the communication medium.

The sources of noise which are in the transmitter and receiver
can be predicted by calculations in many situations. However, noise,
which comes from the channel, is normally more unpredictable since
it may be caused by such things as lightning, electrical machinery
and cosmic disturbances.

In this chapter, we will look at the various modulation schemes
in regards to the signal-to-noise ratio at the output of the receiver's
detector. This seems to be simpler than it really is since it is difficult to compare modulation methods unless all properties of the
system are similar. Our analysis will be based on the assumption
that the noise sources are white in their frequency makeup and the
modulation signal is sinusoidal in nature.

DOUBLE SIDEBAND FULL CARRIER

Refer to Fig. 9-1. We see an amplitude modulation transmit-

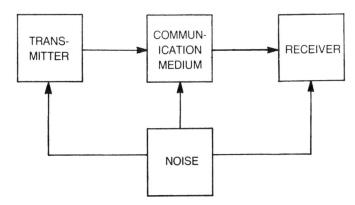

Fig. 9-1. Model of a typical communication system.

ter and a superheterodyne receiver. The transmitter produces a signal of the form given in Equation 9-1.

Equation 9-1

$$e_{cm} = E_c (1 + m\cos\omega_m t) (\cos \omega_c t)$$

where e_{CM} is the instantaneous modulated carrier signal,
E_c is the carrier amplitude,
m is the modulation index,
$\cos \omega_m t$ is the normalized modulation signal,
$\cos \omega_c t$ is the normalized carrier signal,
ω_m is the radian modulation frequency,
ω_c is the radian carrier frequency.

If we expand the signal given in Equation 9-1 by the identity given in Equation 9-2, we obtain Equation 9-3.

$$(\cos A) (\cos B) = \frac{\cos (A + B) + \cos (A - B)}{2}$$

Equation 9-2

$$e_{CM} = E_c\cos \omega_{ct} + mE_c/2 \cos (\omega_c + \omega_m)t + mE_c/2 \cos (\omega_c - \omega_m)t$$

Equation 9-3

Equation 9-3 shows that there are three signals in the wave. These are summarized below:

$E_c \cos \omega_{ct}$ is the carrier signal.

$mE_c/2 \cos (\omega_c + \omega_m)t$ is the upper side frequency signal.

$mE_c/2 \cos (\omega_c - \omega_m)t$ is the lower side frequency signal.

This waveform for Equation 9-3 is shown in the frequency domain by Fig. 9-2. We notice that the carrier amplitude is E_c and each side frequency has an amplitude of $mE_c/2$.

As this amplitude modulated signal is transmitted through the communication system channel, it goes under attenuation and then is intercepted by the receiver antenna. The signal is amplified by the rf amplifier in the receiver and passed on to the mixer circuit. It is beat with the local oscillator signal to produce the intermediate frequency signal or i-f signal. This signal is amplified by the i-f amplifier and presented to the AM detector. The signal at the detector input appears similar to the signal given by Equation 9-3. This signal can be written as Equation 9-4.

Equation 9-4

$$e_{i\text{-}f} = E_{i\text{-}f} \cos \omega_{i\text{-}f} \, t \quad + \quad mE_{i\text{-}f}/2 \cos (\omega_{i\text{-}f} + \omega_m) \, t$$
$$+ \quad mE_{i\text{-}f}/2 \cos (\omega_{i\text{-}f} - \omega_m)t$$

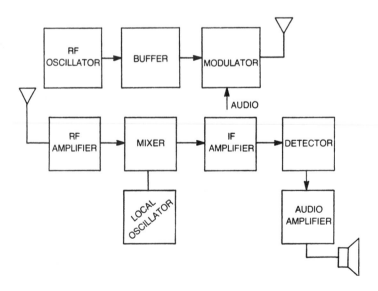

Fig. 9-2. Amplitude modulation transmitter and a superheterodyne receiver.

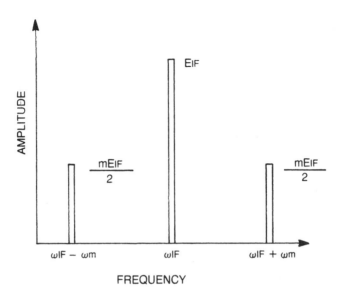

Fig. 9-3. Frequency domain for double sideband full carrier signal.

The frequency spectrum for this signal is gives by Fig. 9-3.

For the process of demodulation, we will assume an envelope detector is present. An example of such a detector is shown in Fig. 9-4. The envelope of the intermediate frequency signal is seen by looking at Equation 9-5, which is equivalent to Equation 9-4.

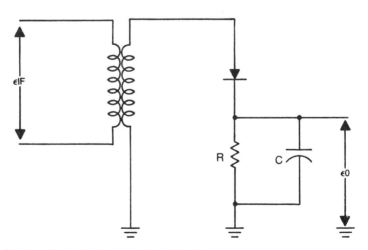

Fig. 9-4. Example of an envelope detector.

Equation 9-5

$$e_{i\text{-}f} = E_{i\text{-}f} (1 + m \cos \omega_m t) (\cos \omega_{i\text{-}f} t)$$

The envelope of the waveform is the portion given by Equation 9-6.

Equation 9-6

$$Envelope = E_{i\text{-}f} (1 + m \cos \omega_m t)$$

To see exactly what the entire signal $e_{i\text{-}f}$ looks like, refer to Fig. 9-5. The envelope has an appearance like the modulation signal.

The total average power present at the input of the receiver is found by summing the average power due to the carrier and the two sidebands at the input to the receiver.

The formula to calculate the average power of any periodic waveform is given by Equation 9-7.

Equation 9-7

$$P = \frac{1}{T} \int_0^T \frac{[v_{(t)}]^2 \, dt}{R}$$

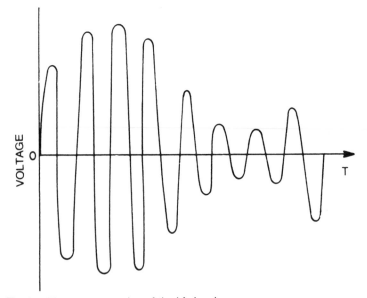

Fig. 9-5. Time representation of the i-f signal.

where P is the average power.

$v_{(t)}$ is the voltage as a function of time.
R is the resistance which has $v_{(t)}$ across it.

We could also use this formula:

Equation 9-8

$$P = \frac{1}{T} \int_0^T \frac{[i_{(t)}]^2 \, R \, dt}{R}$$

where P is the average power.

$i_{(t)}$ is the current as a function of time.

R is the resistance which has $i_{(t)}$ flowing through it.

For a carrier waveform of $E_c \sin \omega ct$, the average power becomes

$$P = \frac{1}{T} \int_0^T \frac{(E_c \sin \omega ct)^2}{R} \, dt$$

$$= \frac{E_c^{\,2}}{TR} \int_0^T \sin^2 \omega ct \, dt$$

$$= \frac{E_c^{\,2}}{T R} \int_0^T \left[\frac{1 - \cos 2 \, \omega ct}{2} \right] dt$$

$$= \frac{E_c^{\,2}}{2TR} \int_0^T dt - \frac{Ec^2}{2 \, TR} \int_0^T \cos 2\omega ct \, dt$$

$$= \frac{E_c^{\,2}}{2 \, TR} + \left| \begin{matrix} t \\ 0 \end{matrix} \right. - \frac{E_c^{\,2}}{2 \, TR} \left[\frac{\sin 2 \, \omega ct}{2 \, \omega c} \right]_0^T$$

$$= \frac{E_c^2}{2R}(T-0) - \frac{E_c^2}{2TR}\left[\frac{\sin 2 \, \omega c \, t}{2 \, \omega c}\right]$$

$$+ \frac{E_c^2}{2TR}\left[\frac{\sin 2 \, \omega c \,(0)}{2 \, \omega c}\right]$$

$$P = \frac{E_c^2}{2R} - \frac{E_c^2}{2TR}\left[\frac{\sin 2\left[\frac{2\pi}{T}\right]T}{2 \, \omega c}\right]$$

$$+ \frac{E_c^2}{2TR}\left[\frac{-\sin 2\left[\frac{2\pi}{T}\right]0}{2 \, \omega c}\right]$$

$$= \frac{E_c^2}{2R} - \frac{E_c^2}{2TR}\left[\sin 4\pi - \sin 0\right]$$

$$= \frac{E_c^2}{2R} - \frac{E_c^2}{2TR}\left[0 - 0\right]$$

$$P = \frac{E_c^2}{2R}$$

Now this can be simplified by Equation 9-9.

Equation 9-9

$$P = \frac{Ec}{\sqrt{2}}^2\left[\frac{1}{R}\right]$$

This simply says that we must take the rms value of the wave voltage square it, and divide by R. This only works for sine and cosine waves. The rms value is different for other waves and of course gives different results (unless by some coincidence).

Had we dealt with current, we could have shown that the average power would come out to be that of Equation 9-10.

Equation 9-10

$$P_{av} = \frac{I_c^2 R}{2}$$

The total power associated with the AM wave received could be written as Equation 9-11.

Equation 9-11

$$P_t = S_i = P_c + P_{USB} + P_{LSB}$$

Because the carrier and sideband components are all sinusoidal waveforms, we could write the following Equation 9-12.

$$S_i = \frac{\left[\dfrac{E_c}{\sqrt{2}}\right]^2}{R} + \frac{\left[\dfrac{mE_c}{2\sqrt{2}}\right]^2}{R} + \frac{\left[\dfrac{mE_c}{2\sqrt{2}}\right]^2}{R}$$

$$= \frac{E_c^2}{2R} + \frac{m^2 E_c^2}{8R} + \frac{m^2 E_c^2}{8R}$$

Equation 9-12

$$= \frac{E_c^2}{2R}\left[1 + \frac{m^2}{2}\right]$$

By a similar evaluation we can find the i-f component average power at the input of the detector to be given by Equation 9-13.

Equation 9-13

$$P_{i\text{-}f} = \frac{E_{i\text{-}f}^2}{2R}$$

Likewise the total average power at the input of the detector to be given by Equation 9-14.

Equation 9-14

$$S_i = \frac{E_{i\text{-}f}^2}{2R}\left[1 + \frac{m^2}{2}\right]$$

If we assume S_i in reference to a 1 ohm resistance we would obtain Equation 9-15.

Equation 9-15

$$S_i = \frac{E_{i\text{-}f}^{\,2}}{2}\left[1 + \frac{m^2}{2}\right]$$

Assuming that white noise is present at the input of the envelope detector, with a spectral density of n, then the total noise power at the input will be given by Equation 9-16.

Equation 9-16

$$N_i = 2\,n\,B$$

Dividing Equation 9-15 by Equation 9-16, we obtain the input signal-to-noise ratio given in Equation 9-17.

Equation 9-17

$$\frac{S_i}{N_i} = \frac{\dfrac{E_{i\text{-}f}^{\,2}}{2}\left(1 + \dfrac{m^2}{2}\right)}{2\,n\,B}$$

$$= \frac{E_{i\text{-}f}^{\,2}\left(1 + \dfrac{m^2}{2}\right)}{4\,n\,B}$$

Each noise component within the bandwidth of the i-f signal will beat with the i-f carrier. The noise components that exist above the i-f carrier to $f_c + B$ and to $f_c - B$ will have identical variances and give rise to a total noise power of $2\,n\,B$ at the output of the envelope detector. As can be seen, this is equivalent to the input noise power given by Equation 9-16.

The power associated with the output signal from the envelope detector is found in the following manner:

The envelope is given by the equation 9-18.

Equation 9-18

$$e = E_{i\text{-}f}\,(1 + m\,\cos\,\omega m t)$$

If we expand this we obtain Equation 9-19.

Equation 9-19

$$e = E_{i\text{-}f} + m\,E_{i\text{-}f}\,\cos\,\omega m t$$

The $E_{i\text{-}f}$ term can be blocked by a capacitor in the detector giving an output as shown by Equation 9-20.

Equation 9-20

$$e' = m\,E_{i\text{-}f}\,\cos\,\omega m t$$

If we find the average power associated with this signal, we obtain Equation 9-21.

Equation 9-21

$$S_0 = \frac{(m\,E_{i\text{-}f})^2}{2} = \frac{m^2\,E_{i\text{-}f}^{\,2}}{2}$$

The output signal-to-noise ratio then is given by Equation 9-22.

Equation 9-22

$$\frac{S_0}{N_0} = \frac{\dfrac{m^2\,E_{i\text{-}f}^{\,2}}{2}}{2\,n\,B} = \frac{m^2\,E_{i\text{-}f}^{\,2}}{4\,n\,B}$$

Making a comparison of the input and output signal-to-noise ratios we have Equation 9-23.

$$\frac{\dfrac{S_i}{N_i}}{\dfrac{S_0}{N_0}} = \frac{\dfrac{E_{i\text{-}f}^{\,2}\,(1 + m^2/2)}{4\,n\,B}}{\dfrac{m^2 E_{i\text{-}f}^{\,2}}{4\,n\,B}} = \frac{1 + m^2/2}{m^2}$$

or,

Equation 9-23

$$\frac{\dfrac{S_i}{N_i}}{\dfrac{S_o}{N_o}} = 1/m^2 + 1/2$$

Example 9-1

The input signal-to-noise ratio for an envelope detector is 100. The modulation index for the incoming AM wave is 0.5. Compute the output signal-to-noise ratio.

$$\frac{S_o}{N_o} = \frac{S_i/N_i}{1/2m^2 + 1/2} = \frac{\dfrac{S_i}{N_i}}{\dfrac{2 + m^2}{2m^2}}$$

$$= \left(\frac{2m^2}{2 + m^2}\right)\frac{(S_i)}{N_i} = \left(\frac{2 \times (.5)^2}{2 + (.5)^2}\right)(100)$$

$$= \frac{0.5}{2.25}(100) \quad = 22.2.$$

Example 9-2

What is the maximum improvement in signal-to-noise ratio at the output that can occur for regular AM with small noise amounts?

$$\frac{S_o}{N_o} = \frac{S_i}{N_i}\left(\frac{(2m^2)}{2 + m^2.}\right)$$

By inspection of the above relation it can be seen that when m is zero, So/No is zero. But when m is unity, which is the highest modulation index allowed, the output signal-to-noise ratio is two-thirds that of the input signal-to-noise ratio.

The case for large noise gives a different solution to the noise problem. Large noise is defined as the situation when the sum of the carrier amplitude and the modulating signal are always smaller than the noise time signal. It is possible that if the noise level is large enough, the envelope detector will completely mutilate the signal. This results in a so-called "threshold" level existing in envelope detectors. The threshold is the value of signal-to-noise ratio below which the output signal-to-noise ratio degrades more quickly than the input signal-to-noise ratio. The threshold effect is first noticed when the input signal-to-noise ratio reaches the vicinity of unity.

DOUBLE SIDEBAND SUPPRESSED CARRIER

In this modulation method, the two sidebands are transmitted but the carrier is not. The signal that is received at the antenna is of the form given by Equation 9-24.

Equation 9-24

$$e = E \, (\cos \omega_m t) \, (\cos \omega_c t)$$

If we expand this formula we obtain Equation 9-25.

Equation 9-25

$$e = E/2 \cos (\omega_c + \omega_m)t + E/2 \cos (\omega_c - \omega_m)t$$

The average power for this waveform referenced to a 1 Ω resistance is the sum of the powers due to the lower and upper side frequencies. This is given by Equation 9-26.

$$S_i = P_{USB} + P_{LSB}$$

$$S_i = (E/2\sqrt{2})^2 + (E/2\sqrt{2})^2$$

$$S_i = E^2/4$$

Equation 9-26

To detect the DSBSC signal we need a system as shown in Fig. 9-6. The detector is referred to as a synchronous detector since the incoming waveform is beat with a carrier of the same phase as that in the waveform. The result is given by multiplying the DSBSC waveform by cos ωct. This is shown in Equation 9-27.

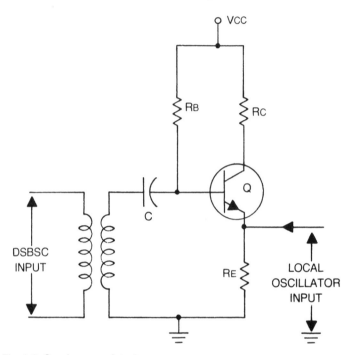

Fig. 9-6. Synchronous detector.

Equation 9-27

$$e_o = (e_{DSBSC}) \text{ (carrier)}$$
$$= E \, (\cos \omega_m t) \, (\cos \omega_c t) \, (\cos \omega_c t)$$
$$= E \, (\cos \omega_m t) \, (\cos \omega_c t)^2$$

Equation 9-28

$$E/2 \cos \omega_m t \ + \ E/2 \, (\cos 2\omega_c t) \, (\cos \omega_m t)$$

If we filter out the high frequency signal $E/2 \cos 2\omega_{ct}$, we obtain Equation 9-29.

Equation 9-29

$$e_o = E/2 \cos \omega_m t$$

The output signal power is given by Equation 9-30.

Equation 9-30

$$S_o = (E/2\sqrt{2^2}) = E^2/8$$

As indicated in the case for small noise when we discussed double sideband full carrier, the noise at the input of the detector has a noise power of $N_i = 2Bn$. We may consider each frequency component in the noise as being $\cos \omega_n t$. Each one of these components beat with the carrier, at the detector, and give rise to a noise power in the output of one-fourth the input noise or $N_i/4$. We then can calculate the input to output signal-to-noise ratio as given by Equation 9-31.

Equation 9-31

$$\dfrac{\dfrac{S_i}{N_i}}{\dfrac{S_o}{N_o}} = \dfrac{\dfrac{E^2}{4}}{\dfrac{N_i}{E^2}} = \dfrac{\dfrac{E^2}{4N_i}}{\dfrac{E^2}{2\,N_i}} = 1/2$$

We can therefore see that the signal-to-noise ratio at the output becomes twice that of the signal-to-noise ratio at the input.

Example 9-3

The input signal-to-noise ratio of a synchronous detector is 100. Calculate the output signal-to-noise ratio.

$$\dfrac{\dfrac{S_i}{N_i}}{\dfrac{S_o}{N_o}} = 1/2 \qquad\qquad \dfrac{S_o}{N_o} = S_i/N_i(2)$$

$$\dfrac{S_o}{N_o} = (100)\,(2) \qquad\qquad \dfrac{S_o}{N_o} = 200.$$

SINGLE SIDEBAND

In a single sideband system either upper or lower side frequencies are transmitted. As an example the formula for an upper frequency is given by Equation 9-32.

Equation 9-32

$$e = E \cos (\omega_c + \omega_m)t$$

This waveform is detected by a synchronous detector. The power at the detector input referenced to 1 ohm is given by Equation 9-33.

Equation 9-33

$$S_i = (E/\sqrt{2})^2 = E^2/2$$

In the detector, the incoming waveform is multiplied by $\cos \omega_c t$. The result is given below by Equation 9-34.

$$e_o = (E \cos(\omega_c + \omega_m t) \cos \omega_c t$$

Equation 9-34

$$e_o = E/2 \cos \omega_m t + E/2 \cos (2\omega_c + \omega_m)t$$

After filtering out the higher frequency component $2 \omega_c + \omega_m$, we get Equation 9-35. This complete operation is shown in Fig. 9-7.

Equation 9-35

$$e_o = E \cos \omega_m t$$

The outside power is given by Equation 9-36.

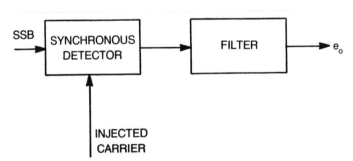

Fig. 9-7. Recovery of modulation from SSB waveform.

Equation 9-36

$$S_o = (E/2\sqrt{2})^2 = E^2/8$$

As for the DSBSC case, the output noise becomes one-fourth the input noise to the detector. The output to input signal-to-noise ratio then becomes Equation 9-37.

$$\dfrac{\dfrac{S_i}{N_i}}{\dfrac{S_o}{N_o}} = \dfrac{\dfrac{E^2}{2}}{\dfrac{E^2}{8}} = 1$$

or,

Equation 9-37

$$\dfrac{\dfrac{S_o}{N_o}}{\dfrac{S_i}{N_i}} = 1$$

There is no change from input to output in signal-to-noise ratio in this situation.

FREQUENCY MODULATION

In frequency modulation, the size of the output signal coming from the detector depends on the amount of frequency deviation produced during modulation. Likewise, this deviation is a direct function of the modulation signal amplitude. Also, noise present in the FM system causes frequency deviation.

It is beyond the scope of this book to make a detailed derivation of how noise effects the FM process but we can say the following:

- The signal power in the output depends on the discrimina-

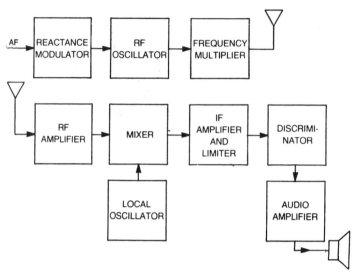

Fig. 9-8. Basic FM system.

tor's ability to convert frequency deviation to output voltage.

• The signal power in the output depends on the frequency deviation in the FM waveform.

The basic FM system can be depicted as shown in Fig. 9-8. The FM waveform which is present at the rf amplifier input is of the form given by Equation 9-38.

Equation 9-38

$$e_{CM} = E_c \sin (\omega_c t + m_f \sin \omega_m t)$$

where e_{CM} is the instantaneous modulated carrier signal,

E_c is the carrier amplitude,

m_f is the modulation index or argument,

ω_m is the radian modulation frequency,

ω_c is the radian carrier frequency.

The incoming signal becomes heterodyned down to a new frequency domain as given by Fig. 9-9. The modulation index m_f is the ratio of the frequency deviation to the modulation signal fre-

228

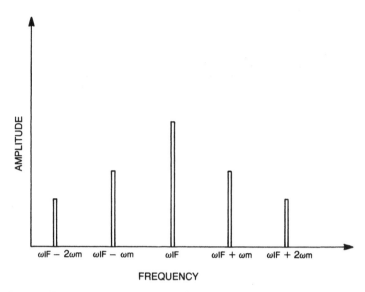

Fig. 9-9. Frequency domain for FM i-f signal.

quency FM. The output of the discriminator is given by Equation 9-39.

Equation 9-39

$$e_o = D \, d\Theta/dt$$

The D is the discriminator constant and dΘ/dt is the derivative of the phase angle for the FM waveform. For the situation we are confronted with, it can be easily shown that Equation 9-40 is true.

Equation 9-40

$$e_o = D \, (\Delta f) \, \sin\omega_m t$$

We notice that the amplitude of this signal is dependent on the product of the discriminator constant and frequency deviation.

The output power from the discriminator referenced to one ohm would be given by Equation 9-41.

Equation 9-41

$$S_o = D^2 \, (\Delta f)^2/2$$

It can be proven by calculus that the noise output of a discriminator increases with the modulation frequency FM, so that the output noise power is given by Equation 9-42.

Equation 9-42

$$N_o = D^2n \ (FM)^2/3 \ Pc$$

The symbol Pc is the carrier power. The signal to noise ratio at the output of the discriminator is derived as follows:

$$\frac{S_o}{N_o} = \frac{\dfrac{D^2 \ (\Delta f)^2}{2}}{\dfrac{D^2 \ n \ (fm)^3}{3 \ Pc}} = \frac{3}{2} \ \frac{(\Delta f)^2}{(fm)^2} \ \frac{Pc}{(fm) \ n}$$

The output signal-to-noise ratio is given by Equation 9-43.

Equation 9-43

$$S_o/N_o = 3/2 \ (m_f)^2 \ Pc/(FM)n$$

or we can write,

$$S_o/N_o = 3 \ (m_f)^2 \times Pc/2 \ n \ (FM)$$

Remembering that the output signal-to-noise ratio for AM was $m^2Ec^2/4nB$ we can write Equation 9-44 in terms of the fact when m equals unity.

$$\frac{S_o}{N_o} = 3 \ (m_f)^2 \ \frac{Pc}{2 \ nB} = 3 \ (m_f)^2 \ \frac{Ec^2/2}{2nB} \bigg|_B$$

$$\frac{S_o}{N_o}\bigg| FM = 3 \ (m_f)^2 \ \frac{Ec^2}{4nB}$$

$$\frac{S_o}{N_o}\bigg| FM = 3 \ (m_f)^2 \ \frac{S_o}{N_o}\bigg| AM$$

Equation 9-44

We can say that an FM system has a signal-to-noise ratio S_o/N_o FM that is 3 $(m_f)^2$ as good as a similar AM system operating with a modulation index of one.

Example 9-4

If the frequency deviation of a system is 12 kHz and the modulating frequency is 4 kHz, what is the improvement due to FM over a comparable AM system with m = 1?

$$\frac{\dfrac{S_o}{N_o}\Big|_{FM}}{\dfrac{S_o}{N_o}\Big|_{AM}} = 3 \, m_f^2 = 3 \, \frac{(\Delta f)^2}{fm^2} = 3\left(\frac{(12k)}{4k}\right)^2 = 27.$$

Example 9-5

What is the maximum improvement in signal-to-noise ratio over the comparable AM system that can occur, when f_m is maximum?

A frequency deviation of 75 kHz is allowed by the FCC with a maximum modulation frequency of 15 kHz. This sets the argument or modulation index at $m_f = \Delta f/f_m = 75$ kHz/15 kHz = 5. The output signal-to-noise ratio of the discriminator is then

$$\frac{S_o}{N_o}\bigg|_{FM} = 3 \, (m_f)^2 \, \frac{S_o}{N_o}\bigg|_{AM}$$

$$= 3 \, (5)^2 \, \frac{S_o}{N_o}\bigg|_{AM}$$

$$= 75 \, \frac{S_o}{N_o}\bigg|_{AM}.$$

Example 9-6

At what value of m_f will an improvement in signal-to-noise ratio be noticed over a comparable AM system with modulation index of one?

As long as 3 $(m_f)^2$ is larger than one, an improvement will be possible. We can therefore solve for m_f.

$$3 \ (m_f)^2 \ = \ 1$$
$$(m_f)^2 \ = \ 1/3$$
$$m_f \ = \ 1/\sqrt{3}$$

The argument should be slightly larger than $1/\sqrt{3}$ for an improvement to be noticed.

In an FM process the modulation index is inversely proportional to the modulation frequency. This means that the modulation index would be greatest for the smaller modulation frequencies while it would be smaller for the larger modulation frequencies. For example, suppose the modulator constant is k_f equals 2000 hertz/volt and the modulation signal has an amplitude of 6 volts.

If the modulation frequency was 1000 hertz, the modulation index would be:

$$m_f \ = \ \frac{k_f \ (E_m)}{f_m}$$

$$= \ \frac{2000 \ (\ 6 \ )}{1000}$$

$$= \ 12.$$

Should the modulation frequency be raised to 12 kHz, the modulation index would be:

$$m_f \ = \ \frac{k_f \ (E_m)}{f_m}$$

$$= \ \frac{2000 \ (6)}{12,000}$$

$$= \ 1$$

Thus, a twelvefold increase in the modulation frequency caused the index to fall by a factor of twelve.

The effective bandwidth for each case is given by Equation 9-45.

Equation 9-45

$$BW = 2 (n) f_m$$

where n is the number of sideband pairs and f_m is the modulation frequency.

It is a known fact that the number of sideband pairs is approximately one more than the modulation index. Therefore, for the case when $f_m = 1$ k and $m_f = 12$, the bandwidth is found as follows:

$$n = 1 + m_f$$
$$= 1 + 12$$
$$= 13$$
$$BW = 2 n f_m$$
$$= 2 (13) (1k)$$
$$= 26 \text{ kHz}$$

For the case when $f_m = 12$ kHz and $m_f =$ unity, the bandwidth is found as follows:

$$n = 1 + mf$$
$$= 1 + 1$$
$$= 2$$
$$BW = 2 (n) (fm)$$
$$= 2 (2) (12 \text{ kHz})$$
$$= 48 \text{ kHz}.$$

Noise on the other hand has an effect on the FM waveform that produces an effective phase type of modulation. Phase modulation is not sensitive to frequency, so the modulation index due to noise remains constant assuming the amplitude of the effective noise remains constant.

To detect an FM waveform a circuit is required to differentiate the waveform. One example of such a circuit is a discriminator. The output of a discriminator looks like the signal that originally caused the modulation. In Fig. 9-10 we see our FM waveform and in Fig. 9-11 we see the derivative of such a wave. Note that the

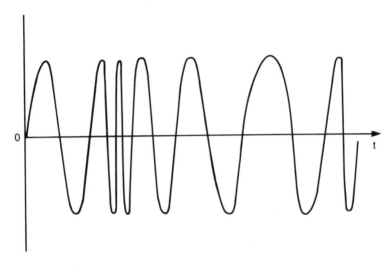

Fig. 9-10. Simple FM waveform.

derivative of the FM waveform increases in amplitude as the deviation increases and decreases in amplitude as the derivation decreases.

We therefore conclude that there is an improvement in signal-to-noise ratio at the output of the FM detector over that of the AM detector. Also, there is less noise at typical FM frequencies which

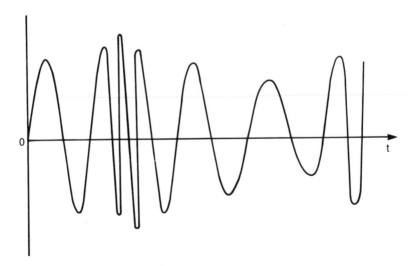

Fig. 9-11. Derivative of an FM waveform.

helps reduce the noise quota. However, because the information content is reflected in the frequency deviation, circuits can be constructed that slice away undesirable amplitude variations and reduce noise even more.

NOISE IN DIGITAL COMMUNICATION SYSTEMS

Before we attack this subject let us discuss briefly the basic idea behind digital communications. In this type of communication system samples of the modulation are taken instead of modulating a carrier by analog means. The samples are taken at a particular rate known as the sampling frequency. These samples may be transmitted directly over a wire, or they can modulate a carrier signal to produce some specific modulation scheme.

Perhaps the most important theorem in digital communications is that a modulation signal can be sampled at a certain rate and still be reconstructed without distortion provided that the rate is at least twice the highest frequency present in the modulation signal. In Fig. 9-12, we see a signal that is composed of only one frequency f_x. If the sampling frequency is f_y, then the resulting frequency spectrum due to the sampling is as shown in Fig. 9-13. We note that if the sampling frequency f_y is exactly twice f_x then the lower side frequency $f_y - f_x$ just touches f_x causing the condition $f_y - f_x = f_x$. This is another way of saying $f_y = 2f_x$. This is called the *Nyquist rate* in digital communications.

In practical situations the modulation has many frequencies in

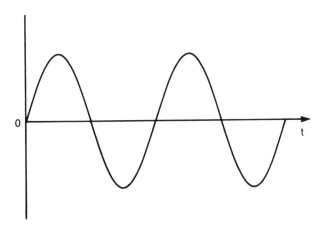

Fig. 9-12. Time spectrum of a single frequency tone.

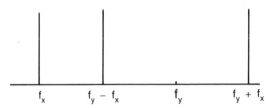

Fig. 9-13. Resulting spectrum if f_y is the sampling frequency.

it. This causes the frequency spectrum to look like that shown in Fig.9-14. If we sample at a frequency less than twice that of the sampling frequency, the lower sideband of the sampled spectrum will intermix with the modulation signal causing distortion.

PAM

In *PAM (pulse-amplitude modulation)* the size of each sample is proportional to the modulation signal size. In Fig. 9-15, we see the result of adding noise to a PAM signal. We are assuming white noise and that the sampling theorem holds, that is $f_y \geq 2\ f_x$. At the time span between samples noise produces additional disturbance power without any signal power increase. We can call the input signal-to-noise ratio to the PAM receiver $\dfrac{S_iP_A}{N_iP_A}$. The evaluation of the process is essentially the same as the case for double sideband suppressed carrier giving a signal-to-noise ratio at the output as given by Equation 9-45.

Equation 9-45

$$\frac{S_oP_A}{N_oP_A} = 2\ \frac{S_iP_A}{N_iP_A}$$

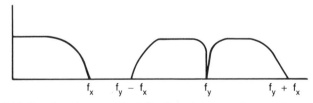

Fig. 9-14. Spectrum for a practical situation when many frequencies are contained in the modulation.

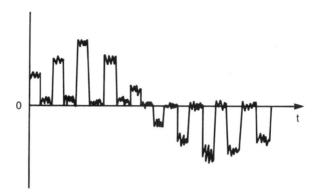

Fig. 9-15. Result of adding noise to a PAM waveform.

PULSE-POSITION MODULATION

In PPM the information is present in the positioning of

a) trailing edge
b) leading edge
c) trailing edge and leading edge

Refer to Fig. 9-16. Here we see the basic PPM signal under the influence of noise. As noise is added to the signal the result is not only a change in amplitude which we will call ΔA but also

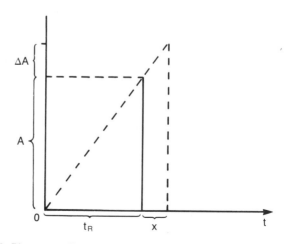

Fig. 9-16. Diagram to illustrate the effect of noise in PPM.

a change in position which we shall call x. If the amplitude is called A and the pulse rise time t_r, we may state Equation 9-46.

Equation 9-46

$$\frac{x}{t_r + x} = \frac{\Delta A}{\Delta A + A}$$

Taking the reciprocal of this we obtain Equation 9-47.

Equation 9-47

$$\frac{t_r + x}{x} = \frac{\Delta A + A}{\Delta A}$$

Dividing, this results in:

$$\frac{t_r}{x} + 1 = 1 + \frac{A}{\Delta A}$$

We may cancel the unity terms giving the following. Taking this reciprocal again we obtain Equation 9-48.

Equation 9-48

$$\frac{x}{t_r} = \frac{\Delta A}{A}$$

This simply says that the ratio of the change in position due to noise to the rise time of the pulse is equal to the ratio of the change in pulse amplitude due to noise in the original pulse amplitude.

We may square both sides of the equation giving Equation 9-49

Equation 9-49

$$\frac{x^2}{t_r^2} = \frac{[\Delta A]^2}{A^2}$$

The noise in the output is then given by Equation 9-50.

Equation 9-50

$$x^2 = \frac{t_r^2 [\Delta A]^2}{A^2}$$

If we assume a maximum displacement due to signal of Δt, we could say that for sinusoidal modulation signal the signal variation has an rms level of $K_{pp} = \dfrac{(\Delta t)}{\sqrt{2^2}}$

The incoming noise is x with a squared value of x^2. The output signal-to-noise ratio is then given as Equation 9-51.

Equation 9-51

$$\frac{S_o}{N_o} = \frac{\left[\dfrac{K_{pp}(\Delta t)}{\sqrt{2}}\right]^2}{K_{pp}^2 \; x^2} = \frac{(K_{pp})^2 \, (\Delta t)^2}{(K_{pp})^2 \, x^2}$$

$$= \frac{(\Delta t)^2}{2\left[\dfrac{t_r^2 [\Delta A]^2}{A^2}\right]}$$

We may rewrite this as in Equation 9-52.

Equation 9-52

$$\frac{S_o}{N_o} = \frac{(\Delta t)^2}{2\,(t_r)^2} \; \frac{A^2}{(\Delta A)^2}$$

But, the input signal-to-noise ratio is given by Equation 9-53.

Equation 9-53

$$\frac{S_i}{N_i} = \frac{A^2}{(\Delta A)^2}$$

Therefore we can write Equation 9-54.

Equation 9-54

$$\frac{S_o}{N_o} = \frac{(\Delta t)^2}{2\,(t_r)^2}\,\frac{S_i}{N_i}$$

The equation states that the output signal-to-noise ratio equals the square of the ratio of the maximum time displacement due to the input signal. The product of twice the rise time squared all multiplied by the input signal-to-noise ratio.

We therefore conclude that the larger the maximum time displacement Δt, the larger the output signal-to-noise ratio of the system. We could write the equation somewhat different by remembering that the rise time t_r is related to the bandwidth by the Equation 9-55.

Equation 9-55

$$B = \frac{k}{t_r}$$

where k is a constant of the system. For the system response, which is a basic low pass characteristic, the rise time is equal to that of Equation 9-56.

Equation 9-56

$$t_r = 2.2\ RC$$

where RC is the time constant of the system.

It is well known that the 3 dB bandwidth of an RC filter is given by Equation 9-57.

Equation 9-57

$$BW = \frac{1}{2\,\pi\,RC}$$

Making a substitution we have

$$t_r = 2.2 \frac{1}{2 \pi B}$$

$$t_r = \frac{0.35}{B}$$

We can then write for the signal-to-noise ratio in respect to bandwidth as Equation 9-59.

Equation 9-59

$$\frac{S_o}{N_o} = \frac{(\Delta t)^2}{2(t_r)^2} \frac{S_i}{N_i} = \frac{(\Delta t)^2}{2 \left(\dfrac{0.35}{B} \right)^2} \frac{S_i}{N_i}$$

$$= \frac{(\Delta t)^2 (B)^2}{2 (0.35)^2} \frac{S_i}{N_i}$$

$$= \frac{(\Delta t)^2 (B)^2}{2 (.1225)} \frac{S_i}{N_i}$$

$$= 4.9 (\Delta t)^2 (B)^2 \frac{S_i}{N_i}$$

This states that the output signal-to-noise ratio is proportional to the bandwidth squared for the case of PPM system with typical low-pass channel characteristics.

PULSE-CODE MODULATION (PCM)

In this type of system, the modulation signal is sampled and then a specific level assigned to each sample. This process is called *quantizing*. After this, the quantizing levels are encoded into a pulse code. Then the code may be transmitted directly over the communication link or the code may be used in a modulation scheme such

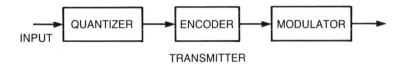

INPUT

TRANSMITTER

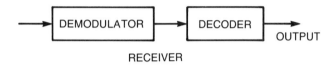

RECEIVER

OUTPUT

Fig. 9-17. Basic PCM System.

as frequency-shift or phase-shift keying. See Fig. 9-17 for a typical transmission system for PCM.

At the receiver the code is removed from the modulation scheme and then decoded. Then the modulation signal may be reconstructed. The overall result is that the receiver must be able to interpret codes values (logic ones and zeros) instead of the shape and/or size of the pulse.

The most important source of noise is referred to as *quantizing noise*. Refer to Fig. 9-18 and note that the quantized signal is made of a group of steps. Because the original signal was not com-

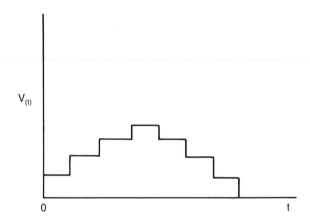

Fig. 9-18. Quantized signal.

posed of a group of steps, there is definitely an error between the two signals.

If we increase the number of steps and the overall range of the original signal remains the same, the error or quantizing noise will be decreased. At the same time, more levels would now be required and more pulses required because a higher number of codes would exist. This would simply mean that the bandwidth must be made larger.

Refer to Fig. 9-19. The range of the signal is shown. Notice that the distance between two quantizing levels is V_q or between

$$- \frac{V_q}{2} \text{ and } \frac{V_q}{2} .$$

To evaluate the rms value of this quantizing noise, we first determine the probability density function as shown in Fig. 9-20. Because the range is between $\frac{V_q}{2}$ and $-\frac{V_q}{2}$ and we know that the area under the curve must be unity, the probability density function must be $\frac{1}{V_q}$.

The variance of the quantizing noise becomes that given by Equations 9-60 and 9-61.

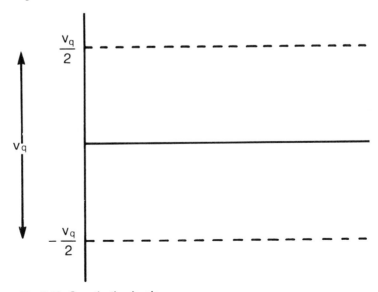

Fig. 9-19. Quantization levels.

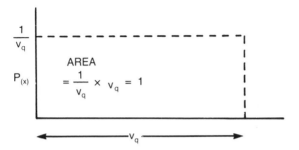

Fig. 9-20. Quantization noise probability curve.

Equation 9-60

$$v_q^2 = \int_{-v_q/2}^{v_q/2} v^2 \, p\,(v_q) \, dv$$

$$= \int_{-v_q/2}^{v_q/2} v^2 \left(\frac{1}{v_q}\right) dv$$

$$= \left[\frac{v^3}{3\,v_q}\right]_{-\frac{v_q}{2}}^{\frac{v_q}{2}}$$

$$= \left[\frac{v_q^{\,3}}{24\,v_q} - \left[\frac{-v_q^{\,3}}{24\,v_q}\right]\right]$$

Equation 9-61

$$v_q^2 = \frac{v_q^2}{12}$$

The rms value is then found by extracting the square root giving Equation 9-62.

Equation 9-62

$$v_{qn} = \frac{v_q}{\sqrt{12}}$$

To obtain a feeling for the output signal-to-noise ratio of such a system, we may think of the signal range as being V and the average normalized power as $\frac{V^2}{2}$. The quantizing level V_q is given by $\frac{V}{2^n}$ since there may be 2^n quantizing levels.

The total noise N_o can be found by Equation 9-63.

Equation 9-63

$$N_o = \frac{\left(\frac{2V}{2^n}\right)^2}{12}$$

where V is the peak-to-peak value of the input waveform.

The resulting signal-to-noise ratio at the output would be found by Equation 9-64.

$$\frac{S_o}{N_o} = \frac{\dfrac{V^2}{2}}{\dfrac{\left(\dfrac{2V}{2^n}\right)^2}{12}}$$

$$= \frac{V^2}{2} \times \frac{12}{\left(\dfrac{2V}{2^n}\right)^2}$$

$$= \frac{V^2}{2} \times \frac{12}{\dfrac{4V^2}{2^{2n}}}$$

$$= \frac{6\,(2^{2n})}{4}$$

Equation 9-64

$$\frac{S_o}{N_o} = (1.5)(2^{2n})$$

As the number of levels are increased, the signal-to-noise ratio in the output increases as shown in Fig. 9-21.

SUMMARY

In this chapter we have seen the various AM communication system signal-to-noise ratios present at the output of the receiver's detector. There are many phenomenons which can affect our calculations in a practical circuit that we did not take into account in this chapter due to their complexity of prediction. However, we have developed signal-to-noise ratios that can be used to approximate and identify a detector circuit's operation in the presence of white noise.

PROBLEMS AND QUESTIONS

1. A DSBFC signal is impressed on an envelope detector. The output signal-to-noise ratio is 200 and the modulation index for the incoming wave is 0.3. Compute the input signal-to-noise ratio.

2. Suppose you were to make an approximate calculation for the noise figure of the envelope detector given in problem one. What formula would you use? What would the noise figure be?

n	$\frac{S_o}{N_o}$
1	6
2	24
3	192

Fig. 9-21. Signal-to-noise ratio versus quantization levels.

3. The input signal-to-noise ratio of a synchronous detector is 67. Calculate the approximate output signal-to-noise voltage.

4. Explain why a synchronous detector produces a noise output which is reduced by a factor of four for the case of double and single sideband.

5. An FM system has a frequency deviation of 25 kHz and a modulation frequency of 15 kHz. What is the output signal-to-noise ratio improvement over a comparable AM system with a modulation index of unity?

6. Suppose the modulation signal for the FM waveform in problem 5 was doubled in amplitude. Repeat problem 5 for this condition. Hint: Frequency deviation is directly proportional to modulation signal amplitude.

7. Research Problem: Explain why the output signal-to-noise ratio increases as the modulation index increases. Hint: the sidebands in an FM wave are dependent on the modulation index.

8. Calculate the rms value of quantizing noise for a PCM system with a quantizing step size of 2 volts.

Chapter 10

Low Noise Circuit Design

I N THE PAST NINE CHAPTERS WE HAVE DISCUSSED MANY TOP-
ics ICS regarding electrical noise. We will now attempt to use
these topics as they relate to low noise circuit design. It must be
remembered, however, that this is a very broad subject. This chapter will be presented in such a way that we become introduced to,
and appreciate, low noise design. We will only touch on a few of
the more interesting topics in this area.

HOW TO WE PREDICT AMPLIFIER NOISE?

As discussed in chapter four, an electronic device such as a
bipolar transistor or field effect transistor has a noise equivalent
resistance. The smaller this resistance, the less noise produced. In
Fig. 10-1 we see a basic voltage amplifier using an FET. Source
bias is produced by direct current flowing through R, the source
resistor. Resistor R_L is the drain load resistor and R_g is the gate
regulation resistor. The source is represented by generator V_g. To
predict the size of the output noise voltage V_{no}, we draw the
equivalent noise circuit for the system as shown in Fig. 10-2A. The
FET is represented by its equivalent circuit along with the noise
equivalent resistance R_{eg}. Since we are interested in only noise
voltage, we may ignore the signal generator voltage V_g. The dynamic resistance of the FET is r_d and the amplification factor of

248

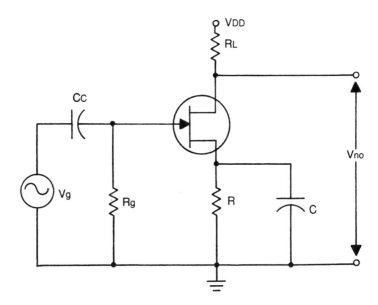

Fig. 10-1. Basic voltage amplifier using a FET.

the FET is μ. The noise voltage e_{ng} is given by Equation 10-1. Note that e_{ng} is caused by the total resistance seen between gate and ground. This assumes the gate to source resistance of the FET is much larger than the sum of the parallel combination of R_s and R_g, added to R_{eg}. The bandwidth of interest is represented by B.

Equation 10-1

$$e_{ng}^2 = 4k\,T\,[(R_s \bullet R_g)/(R_s + R_g) + (R_{eq})](B)$$

At the output terminals (drain to ground) exists the output noise of the stage; V_{no}. The simplest way to calculate the output noise is to calculate V_{no}^2 and then extract the square root. Equation 10-2 represents this operation.

Equation 10-2

$$V_{no}^2 = e_{ng}^2\,A_v^2 + 4\,kT\,R_L\,B$$

We note that e_{ng}^2 must be multiplied by the square of the voltage gain and added to the square of the rms noise voltage due to R_L.

249

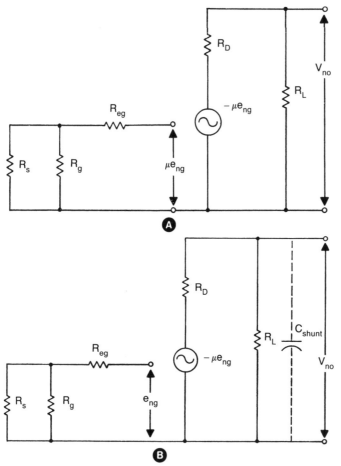

Fig. 10-2(A). Equivalent noise circuit for the system shown in Fig. 10-1. (B) Equivalent circuit for Fig. 10-1 at high frequencies.

Substituting Equation 10-1 into Equation 10-2 and extracting the square root, we obtain the rms noise voltage at the output which is given by Equation 10-3.

Equation 10-3

$$V_{no} = \sqrt{\frac{4\,kT(B)\,(R_s R_g + R_{eq})\,A_v^2 + 4k\,TR_L B}{R_s + R_g}}$$

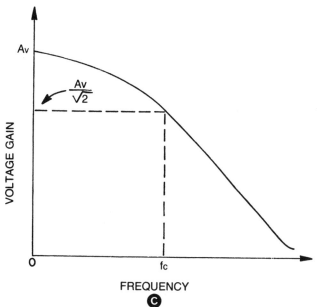

Fig. 10-2(C) Voltage gain versus frequency for the circuit of Fig. 10-1.

The voltage gain for the amplifier stage is given by Equation 10-4.

Equation 10-4

$$A_v = \frac{-\mu Z_L}{Vd + Z_L}$$

In this case, the load impedance Z_L is the load resistor R_L.

If it is desired to calculate the total rms noise output voltage which considers noise contributions at all frequencies, then the noise equivalent bandwidth is needed to make the calculation. For an amplifier as shown in Fig. 10-1, the noise equivalent bandwidth may be calculated in the following manner. First, the equivalent circuit is drawn which determines the voltage gain. This is shown in Fig. 10-2B. At very high frequencies, the shunt capacity of the output circuit must be considered since it determines the cutoff frequency and the noise equivalent bandwidth. The load on the amplifier is R_L in parallel with the shunt capacity as given by Equation 10-5.

Equation 10-5

$$Z_L = \frac{R_L(1/j\omega C)}{R_L + 1/j\omega C} = \frac{R_L}{1 + j\omega\, C\, R_L}$$

Placing Equation 10-5 into Equation 10-4 we obtain the voltage gain as a function of frequency.

$$A_{v(f)} = \frac{-\mu\, Z_L}{r_d + Z_L} = \frac{-\mu(R_L/1 + j\omega C R_L)}{r_d + R_L/1 + j\omega\, C\, R_L}$$

Equation 10-5

$$A_v(f) = \frac{-\mu R_L}{r_d + R_L + j\omega\, C\, R_L\, r_d}$$

If we divide both numerator and denominator by $r_d + R_L$, we obtain Equation 10-6.

Equation 10-6

$$A_{v(f)} = \frac{\dfrac{-\mu\, R_L}{r_d + R_L}}{\dfrac{1 + j\omega\, R_L\, r_d\, C}{r_d + R_L}}$$

This is just a simple transfer function as discussed in chapter two. We may rewrite Equation 10-6 and obtain the voltage gain in terms of the voltage gain at low frequencies and the cutoff frequency. This is given by Equation 10-7.

Equation 10-7

$$A_{v(f)} = \frac{A_v}{1 + j\omega/\omega c}$$

where A_v is the voltage gain at low frequencies, and ω_c is the radian cutoff frequency given by Equation 10-8.

Equation 10-8

$$\omega_c = 2\pi f c$$

where $f_c = (1/2\pi) (R_L + r_d)/R_L \, r_d \, C$

Figure 10-2C shows how the voltage gain varies as frequency changes. We learned in Chapter 5 that the noise equivalent bandwidth of a transfer function, like that given by Equation 10-7, is given by Equation 10-9.

Equation 10-9

$$B_{eq} = (\pi/2) \, (f_c)$$

By substituting the cutoff frequency formula into Equation 10-9, we obtain Equation 10-10.

Equation 10-10

$$B_{eq} = (r_d + R_L)/4 \, R_L \, r_d \, C$$

It is interesting to note that all of the previous information applies also to a triode voltage amplifier.

Example 10-1

Given the voltage amplifier shown in Fig. 10-3, calculate the output noise voltage. The shunt capacity of the output is 100 pf. The FET has a noise equivalent resistance of 1000 Ω.

1. Calculate e_{ng}^2 and assume a temperature of 290 degrees Kelvin. Use Equations 10-1 and 10-10. The noise equivalent bandwidth is calculated by Equation 10-10.

$$B_{eq} = \frac{10 \times 10^3 + 10 \times 10^3}{4 \times 10 \times 10^3 \times 10 \times 10^3 \times 100 \times 10^{-12}} = 500 \text{ kHz}$$

$$e_{ng}^2 = 16 \times 10^{-12} \text{ volts}^2$$

We have made the approximation that the parallel combination of

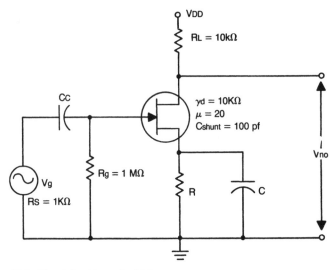

Fig. 10-3. Circuit for Example 10-1.

R_s and R_g is nearly R_s since R_g is much larger than R_s. Therefore 1000 Ω in parallel with 1 megohm is essentially 1000 Ω.

2. Calculate the noise due to R_L.

$$4kTR_LB_{eq} = 4 \times 1.38 \times 10^{-23} \times 2.9 \times 10^2 \times 10^4 \times 5 \times 10^5$$

$$= 80 \times 10^{-12} \text{ volts}^2$$

3. Calculate the voltage gain at low frequencies.

$$A_v = \frac{-\mu R_L}{r_d + R_L} = \frac{-20 \times 10 \times 10^3}{10 \times 10^3 + 10 \times 10^3} = -10$$

4. Calculate V_{no}^2

$$V_{no}^2 = e_{ng}^2 A_v^2 + 4kTR_LB_{eq}$$

$$= 16 \times 10^{-12} (10^2) + 80 \times 10^{-12}$$

$$= 16 \times 10^{-12} \times 10^2 + 80 \times 10^{-12}$$

$$= 16 \times 10^{-10} + 80 \times 10^{-12}$$
$$= 16 \times 10^{-10} + .80 \times 10^{-10}$$
$$= 16.8 \times 10^{-10} \text{ volts}^2$$

5. Calculate V_{no}

$$V_{no} = \sqrt{V_{no}^{2}}$$
$$= \sqrt{16.8 \times 10^{-10}}$$
$$V_{no} = 4.1 \times 10^{-5} \text{ volts}$$

or, $V_{no} = 41 \ \mu V$.

We notice that the majority of the noise voltage squared was determined by the noise present at the gate. Only a small portion was due to the noise from R_L. We note that the higher the voltage gain of the stage, the less important is the noise contribution due to R_L. For example, had the voltage gain of the FET amplifier been -20 due to an amplification factor of -40, then the output noise voltage squared would have been 64.8×10^{-10} volts^2, of which only 0.8×10^{-10} volts^2 were due to R_L. One way of lowering the output noise for this circuit is to use an FET that has a low noise equivalent resistance. However, a problem exists since both the noise equivalent resistance and amplification factor depend on gm. We know that the noise equivalent resistance is inversely proportional to the transconductance, but the gain is directly proportional to the transconductance. This is true since the voltage gain is directly proportional to the amplification factor and the $\mu = g_m \ r_d$. Therefore we must make a compromise between low noise and low voltage gain or high noise and high voltage gain.

DETERMINATION OF
NOISE FIGURE FOR A VOLTAGE AMPLIFIER

In the previous discussion we learned how the noise output voltage is computed for a voltage amplifier. Now let us learn how the noise figure is calculated for that stage.

Refer to Fig. 10-4 which shows the input circuit for the amplifier. We will assume that all capacitors are acting as shorts at the operating frequency of the amplifier. In Chapter 6 we learned the formula for noise figure which is repeated here:

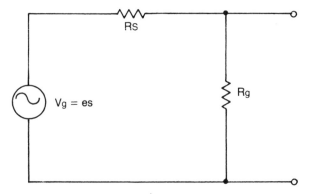

Fig. 10-4. Input circuit for a common source voltage amplifier.

$$F = \frac{\dfrac{S_i}{N_i}}{\dfrac{S_o}{N_o}}$$

The input signal-to-noise ratio as calculated for this type of circuit is given by Equation 10-11.

Equation 10-11

$$S_i/N_i = e_s^2/4kTR_sB_{eq}$$

Symbol e_s is the rms open circuit voltage of the generator.

To determine the output signal-to-noise ratio we must calculate the output signal power S_o and noise power N_o. The amount of open circuit signal generator voltage that actually appears between gate and ground is given by Equation 10-12.

Equation 10-12

$$e = \frac{e_s \times R_g}{R_s + R_g}$$

The output voltage across R_L will be given by Equation 10-13.

Equation 10-13

$$e_o = \frac{(e_s \times R_g)(A_v)}{R_s + R_g}$$

To calculate the output power we square e_o and divide by R_L which gives Equation 10-14.

Equation 10-14

$$S_o = \frac{\dfrac{(e_s \times R_g)^2 (A_v)^2}{(R_s + R_g)^2}}{R_L}$$

Next we calculate the noise output power. This is simply the square of the output noise voltage divided by R_L. The square root of the noise output voltage was given by Equation 10-3 which can be rearranged to obtain Equation 10-15.

Equation 10-15

$$V_{no}^2 = 4kTB_{eq}A_v^2 \left(\frac{R_s R_q}{R_s + R_g} + R_{eq} + \frac{R_L}{A_v^2}\right)$$

We have divided R_L by the square of the voltage gain since we can then write R_L/A_v^2 in series with $R_s R_g/(R_s + R_g) + R_{eq}$. The output power is then given by Equation 10-16.

Equation 10-16

$$\frac{V_{ro}^2}{R_L} = \frac{4kTB_{eq}\left(\dfrac{R_s R_g}{R_s + R_g} + R_{eq} + \dfrac{R_L A_v^2}{A_v^2}\right)}{R_L}$$

To obtain the output signal-to-noise ratio we divide Equation 10-13 by 10-16 and obtain Equation 10-17.

Equation 10-17

$$\frac{S_o}{N_o} = \frac{(e_s)^2 \left(\dfrac{R_g}{R_s + (R_g)}\right)^2}{(4kTB_{eq})\left(\dfrac{R_g R_s}{R_s + R_g} + R_{eq} + \dfrac{R_L}{A_v^2}\right)}$$

Dividing Equation 10-11 by 10-17 we find the noise figure to be given by Equation 10-18.

Equation 10-18

$$F = 1 + \frac{R_s}{R_g} + \left(\frac{R_{eq}}{R_s} + \frac{R_L}{R_s A_v^2}\right)\left(\frac{R_s + R_g}{R_g}\right)^2$$

In many circumstances R_g is much larger than R_s. For example, it is not uncommon for a typical voltage amplifier using an FET or vacuum tube to have $R_g = 1000\ R_s$. The noise figure can then be written as in Equation 10-19.

Equation 10-19

$$F = 1\ \frac{R_{eq}}{R_s} + \frac{R_L}{R_s A_v^2}$$

Example 10-2

Calculate the noise figure for the circuit given in Example 10-1. Since $R_g > R_s$, we may use Equation 10-19.

$$F = 1 + \frac{R_{eg}}{R_s} + \frac{R_L}{R_s\ (A_v)^2}$$

$$= 1 + \frac{1\ k\omega}{1\ k\omega} + \frac{10\ k\omega}{1\ k\omega\ (10)^2}$$

$$= 1 + 1 + .1$$

$$F = 2.1.$$

For very large voltage gains we may write Equation 10-20 which is true only if $R_g > > R_s$.

Equation 10-20

$$F = 1 + R_{eq}/R_s$$

NOISE FIGURE FOR COMMON DRAIN CONFIGURATION

In Fig. 10-5 we see a drain follower circuit which has a maximum voltage gain of unity. By using a similar derivation to the one used for the common source circuit, we find the following. The signal-to-noise ratio at the input is given by Equation 10-21A.

Equation 10-21A

$$S_i/N_i = e_s^2/4kTR_s B_{eq}$$

The output power due to the signal is given by Equation 10-21B.

Equation 10-21B

$$S_o = [(e_s R_g)/(R_s + R_g)]^2 \cdot (1/R)$$

This makes the assumption that the voltage gain is unity.

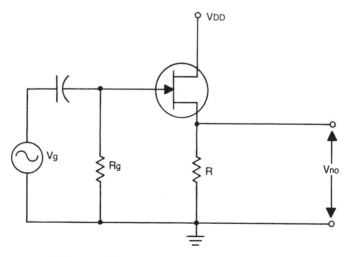

Fig. 10-5. Typical drain follower circuit.

The noise output power is given by Equation 10-22.

Equation 10-22

$$No = 4kTB_{eq}\left(\frac{R_s + R_g}{(R_s R_g)} + R_{eq} + R\right)(1/R)$$

Applying these various formulas in the formula for noise figure we obtain Equation 10-23.

Equation 10-23

$$F = 1 + \frac{R_s}{R_g} + \left(\frac{R_s + R_g}{R_g}\right)^2 \left(\frac{R_{eq}}{R_s} + \frac{R}{R_s}\right)$$

If R_g is much larger than R_s, we may write Equation 10-24.

Equation 10-24

$$F = 1 + R_{eq}/R_s + R/R_s$$

Example 10-3

A common drain circuit has an FET with R_{eq} of 1 kΩ and source resistance of 1 kΩ. If the resistance between the FET source and ground is 5 kΩ, calculate the noise figure. Also $R_g >> R_s$.

$$F = 1 + R_{eq}/R_s + R/R_s$$
$$= 1 + 1 \text{ k}\Omega/1 \text{ k}\Omega + 5 \text{ k}\Omega/1 \text{ k}\Omega$$
$$F = 7.$$

We note that the noise figure of the drain follower is directly related to the resistor connected between the FET source and ground. If this resistor becomes large the noise figure becomes large.

Example 10-4

Repeat Example 10-3 if $R = 40$ kΩ.

$$F = 1 + R_{eq}/R_s + R/R_s$$

$$= 1 + 1 \text{ k}\Omega/1 \text{ k}\Omega + 40 \text{ k}\Omega/1 \text{ k}\Omega$$

$$F = 42.$$

It is not uncommon to find drain followers with large noise figures. This is brought on by the fact that the voltage gain is not larger than unity.

NOISE FIGURE FOR COMMON GATE CONFIGURATION

For the common gate voltage amplifier shown in Fig. 10-6, the noise figure may be calculated by the following procedure.

The signal-to-noise ratio at the input (across R) is given by Equation 10-25.

Equation 10-25

$$S_i/N_i = e^2/4kTR_sB_{eq}$$

Noise power in the output can be written as in Equation 10-26.

Equation 10-26

$$No = [4 \, kTB_{eq} \frac{R_s R}{R_S + R} + R_{eq}) (A_v)^2] /R_L$$

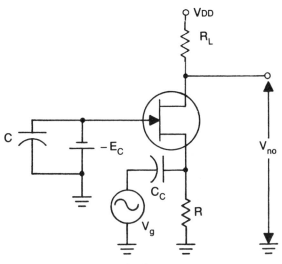

Fig. 10-6. Typical common gate circuit.

The signal output power is given by Equation 10-27A.

Equation 10-27A

$$S_o = e_s^2 (R/R + R_s)^2 (A_v)^2/R_L$$

The voltage gain of a common gate configuration is given by Equation 10-27B.

Equation 10-27B

$$A_v = (\mu + 1) R_L /(r_d + R_L)$$

Application of these formulas gives the formula for noise figure given by Equation 10-28.

Equation 10-28

$$F = 1 + \frac{R_s}{R} + \frac{R_{eq}}{R_s} \left(\frac{R + R_s}{R} \right)^2$$

Example 10-5

Compute the noise figure for the common gate configuration shown in Fig. 10-6.

$$F = 1 + \frac{R_s}{R} + \frac{R_{eq}}{R_s} \left(\frac{R + R_s}{R} \right)^2$$

$$= 1 + 1\,K\,5\,K + 2\,K/1\,K\,(1\,K + 5\,K/5\,K)^2$$

$$= 1 + 0.2 + 2(1.2)^2$$

$$= 1 + 0.2 + 2(1.44)$$

$$F = 1.2 + 2.88$$

$$F = 4.08.$$

MINIMIZATION OF NOISE FIGURE FOR A COMMON SOURCE VOLTAGE AMPLIFIER

If the noise produced by R_L in Fig. 10-4 is negligible, and $R_g >> R_s$, we can write the noise figure for the common source voltage amplifier as Equation 10-29.

Equation 10-29

$$F = 1 + R_s/R_g + R_{eq}/R_s$$

It is noted that if R_s becomes small, R_{eq}/R_s becomes large and R_s/R_g becomes small. If R_s becomes large, R_{eq}/R_s becomes small and R_s/R_g becomes large. Therefore there must be some value of R_s that will give a minimum noise figure. The value of R_s for minimum noise figure is given by Equation 10-30.

Equation 10-30

$$R_{so} = \sqrt{(R_{eq})\,(R_g)}$$

The noise figure then becomes that of Equation 10-30.

Equation 10-30

$$F_{MIN} = 1 + 2\sqrt{R_{eq}/R_g}$$

Example 10-6

For the circuit in Figure 10-3 compute the value of R_s for minimum F. The value of R_g is 1 MΩ and R_{eq} = 1 kΩ.

$$R_{so} = \sqrt{(1 \times 10^3)\,(1 \times 10^6)}$$

$$= \sqrt{10^9}$$

$$= \sqrt{10 \times 10^8} = 31,600$$

The noise figure would then be given by

$$F_{MIN} = 1 + 2\sqrt{R_{eq}/R_g}$$

$$F_{MIN} = 1 + 2\sqrt{1\ \text{k}/1000\ \text{k}}$$

$$= 1 + 2\sqrt{1/1000}$$

$$F_{MIN} = 1 + 2 \times .0316$$

$$F_{MIN} = 1 + .0632$$

$$= 1.0632$$

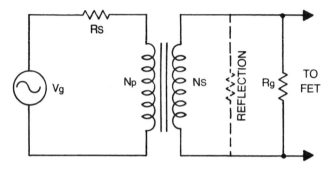

Fig. 10-7. Using a transformer to transform impedance.

Since we need a source resistance of a particular value to result in minimum noise figure, and the generator's resistance R_s is different than R_{so}, the only real solution is to use a transformer to convert impedances. Figure 10-7 shows how this is accomplished. The transformer transformers R_s to R_{so} by way of the turns ratio. The resistance on the primary side of the transformer is multiplied by the square of the turns ratio to give the reflection seen on the secondary side. Equation 10-31 indicates this.

Equation 10-31

$$R_p(N_s/N_p)^2 = R_{sec}$$

where R_p is primary side resistance,

R_{sec} is the reflection of the primary resistance seen on the secondary side, N_p and N_s are the primary and secondary turns respectively.

Since we want to transform R_s to R_{so} we may write Equation 10-32.

$$R_{so} = R_s (N_s/N_p)^2$$

or

Equation 10-32

$$R_{so}/R_s = (N_s/N_p)^2$$

We may also write Equation 10-33.

Equation 10-33

$$(N_s/N_p)^2 = \frac{\sqrt{(R_{eq})\,(R_g)}}{R_s}$$

Example 10-7

What turns ratio should the transformer have for Example 10-6?

$$(N_s/N_p)^2 = R_{so}/R_s = 31{,}600/R_s$$

$$N_s/N_p = \sqrt{31{,}600}/\sqrt{R_s}$$

$$N_s/N_p = 175/\sqrt{R_s}$$

As we can see the ratio would depend on $\sqrt{R_s}$.

NOISE FIGURE MINIMIZATION IN BIPOLAR TRANSISTOR VOLTAGE AMPLIFIER

In Fig. 10-8 is a voltage amplifier using a bipolar transistor in the common emitter configuration. Emitter stability is governed by resistor R_E, and capacitor C_E bypasses the signal around R_E to prevent degeneration. Resistors R_c, R1 and R_s serve for biasing and resistor R_c also acts as the collector load resistor. The output signal is available between collector and ground. The signal generator has an open circuit voltage e_s and internal resistance R_s. Coupling of the source to the amplifier is performed by capacitor C_c.

There are two methods of minimization of the noise figure for this type of circuit—minimization of noise figure by designing for a particular value of collector current and source resistance.

Minimization In Respect To Source Resistance

In this situation, the value of source resistor to use is given by Equation 10-34.

Equation 10-34

$$R_{so} = \sqrt{h_{FE}(1 + 2\,gm\,r_b)/gm}$$

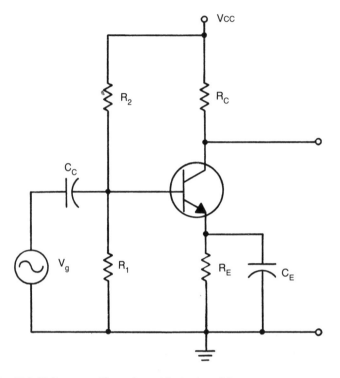

Fig. 10-8. Voltage amplifier using a bipolar transistor.

where h_{FE} is the dc current gain for the transistor, gm is the transconductance and r_b is the base spreading resistance of the transistor.

The noise figure then becomes that given by Equation 10-35.

Equation 10-35

$$F_M = 1 + (R_{so}) \, (gm)/h_{FE}$$

Example 10-8

Compute the minimum noise figure possible with varying R_s if $h_{FE} = 50$, $r_b = 100\Omega$, and $gm = 50 \times 10^{-3}$ V.

$$R_{so} = \frac{\sqrt{50(1 + 2 \times 50 \times 10^{-3} \times 100)}}{50 \times 10^{-3}}$$

$$= \frac{\sqrt{50(1 + 10)}}{50 \times 10^{-3}}$$

$$= \frac{\sqrt{550}}{50 \times 10^{-3}} = 460 \ \Omega$$

$$F_M = 1 + (R_{so}) \ (gm)/h_{FE}$$

$$= 1 + 460 \ (50 \times 10^{-3})/50$$

$$= 1 + .460$$

$$= 1.46$$

Figure 10-9 shows how transformer action could be used to transform the generator's internal resistance to produce the actual optimum resistor and obtain minimum noise figure.

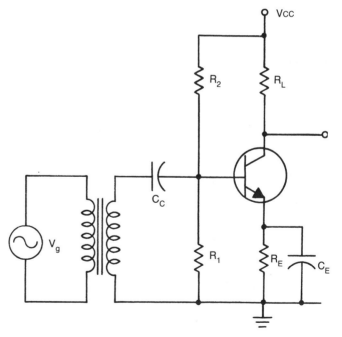

Fig. 10-9. A transformer changes a generator's resistance to produce optimum source resistance for the bipolar transistor voltage amplifier.

It is noted that when dealing with semiconductors, the trans-conductance can be calculated by Equation 10-36.

Equation 10-36

$$gm \cong 40 \, I_c$$

This is valid for 20 degrees Centigrade.

Minimization In Respect To Collector Current

To design for minimum noise figure by using the optimum collector current, we first calculate the collector current for minimum noise figure by using Equation 10-37. This is valid for temperatures near 20 degrees Celsius.

Equation 10-37

$$I_{co} = \frac{\sqrt{h_{FE}}}{40 \, R_s}$$

The minimum noise figure for this collector current is calculated using Equation 10-38.

Equation 10-38

$$F_M = 1 + r_b/R_s + 1/\sqrt{h_{FE}}$$

Example 10-9

Compute the minimum noise figure possible with varying collector current. The h_{FE} is 50, $r_B = 100 \, \Omega$ and source resistance is $100 \, \Omega$. What value of collector current should be used?

$$I_{co} = \frac{\sqrt{50}}{40(100)} = \frac{7.07}{4000}$$

$$= 1.75 \text{ mA}.$$

$$F_M = 1 + \frac{100}{100} + \frac{1}{\sqrt{50}}$$

$$= 1 + 1 + \frac{1}{7.07}$$

$$F_M = 2.141.$$

Noise figure versus source resistance and collector current are shown in Fig. 10-10. Note how noise figure first drops reaching a minimum as either R_s and I_c are varied and then increases once again as R_s and I_c continue to be increased. Now let us look at a further problem dealing with minimization of noise figure.

Example 10-9

A transistor voltage amplifier has been designed for minimum

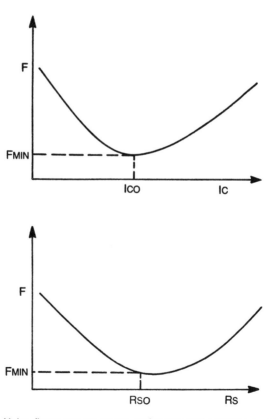

Fig. 10-10. Noise figure versus source resistance and collector current.

noise figure by adjustment of collector current. The collector current is 10 mA and the h_{FE} of the transistor is 100. If the transistor is working at 20 degrees Centigrade, compute the source resistance and transconductance.

$$I_{co} = \sqrt{h_{FE}}/40R_s$$

$$R_s = \frac{\sqrt{h_{FE}}}{40I_{co}} = \frac{\sqrt{100}}{40(10 \times 10^{-3})}$$

$$= 10/.4 = 25\ \Omega$$

$$gm = 40\ I_c$$

$$= 40(10 \times 10^{-3})$$

$$= 400 \times 10^{-3}$$

$$= 400\ m\Omega.$$

NOISE IN INTEGRATED CIRCUITS

Noise is generated in all integrated circuits whether they are analog or digital in nature. Since these circuits are composed of transistors and resistors, various types of noise result.

Analog Integrated Circuits

The operational amplifier is the building block of many types of analog systems. The integrated circuit version of the operational amplifier normally has numerous transistors and resistors. The transistors generate shot and thermal noise while the resistors generate thermal noise.

A noise model can be drawn for an operational amplifier as shown in Fig. 10-11. Such a model shows a noise voltage genera-

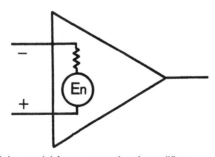

Fig. 10-11. Noise model for an operational amplifier.

tor connected in series with the input resistance.

This noise voltage generator is an equivalent noise source. The generator represents noise voltage for the amplifier over a particular frequency range. A current source of noise may also be shown in parallel with the input resistance. These noise currents and voltages may be stated for a particular bandwidth or as a special density. The spectral density may be stated as follows:

1. volts square per hertz
2. volts per square root of hertz
3. amps squared per hertz
4. amps per square root of hertz

Noise Gain

Whether an operational amplifier is connected in the inverting or noninverting configuration noise generated by the operational amplifier is amplified by a quantity we call the noise gain.

To determine the noise gain refer to Fig. 10-12. We see a basic inverting amplifier configuration. It is known from the theory of how an operational amplifier operates. The voltage gain is given by Equation 10-39.

Equation 10-39

$$A_{CL} = - R_2/R_1$$

where A_{CL} is the closed loop gain of the amplifier circuit. The mi-

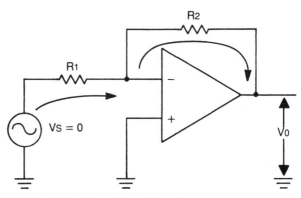

Fig. 10-12. Circuit for determining noise gain.

nus sign stands for the phase reversed of voltage from the input to output.

Now suppose we are interested only in the noise output voltage of the amplifier. We look at the source voltage as if it were not present and its internal resistance were grounded. We draw a current path around the circuit as shown in Fig. 10-12. No current is assumed flowing into the inverting input terminal, since an operational amplifier has a very large input resistance. The current through R1 and R2 is then the same and we may write Equation 10-40.

Equation 10-40

$$\frac{V_s - E_n}{R_1} = \frac{E_n - V_o}{R_2}$$

Setting the source V_s to zero we obtain Equation 10-41.

Equation 10-41

$$\frac{-E_n}{R_1} = \frac{E_n - V_o}{R_2}$$

Solving Equation 11-41 for the noise gain.

$$\frac{E_n}{R_1} - \frac{E_n}{R_2} = -\frac{V_o}{R_2}$$

$$\frac{-E_n}{R_1} - \frac{-E_n}{R_2} = -\frac{V_o}{R_2}$$

$$E_n(1/R_1 + 1/R_2) = V_o/R_2$$

$$\frac{V_o}{E_n} = \frac{R_2}{R_1} + \frac{R_2}{R_2}$$

The noise gain is then given by Equation 10-42.

$$A_n = V_o/E_n = R_2/R_1 + 1$$

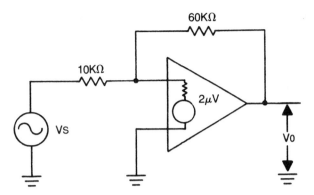

Fig. 10-13. Circuit for Example 10-10.

Example 10-10

A 741 operational amplifier has a total noise voltage of 2 μV of total noise over a frequency range between 10 Hz and 10 kHz. When the source resistance is larger than 30 k, the noise increases. The noise voltage is shown between the inverting and noninverting terminals. Figure 10-13 shows the amplifier circuit.

To calculate the noise output, multiply the equivalent noise voltage by the noise gain.

Equation 10-42

$$V_o = (1 + R_2/R_1)E_n$$

$$= (1 + 60 \text{ k}/10 \text{ k}) (2 \ \mu V)$$

$$= (1 + 6) (2 \ \mu V)$$

$$= 14 \ \mu V.$$

This voltage value is for a bandwidth from 10 Hz to 10 kHz since that was the bandwidth for which E_n was defined.

As was seen in working this problem, the noise gain is 1 + R_2/R_1 = 7. The signal gain would be $-R_2/R_1$ or -6.

NOISE PLOTS FOR OPERATIONAL AMPLIFIERS

Three plots concerning the noise behavior for an operational amplifier are quite typical. They are shown in Fig. 10-14. Note that the noise voltage squared per Hertz and the noise current squared

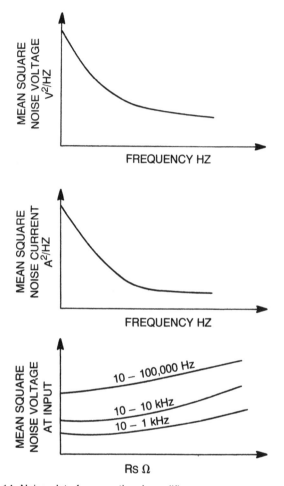

Fig. 10-14. Noise plots for operational amplifiers.

per Hertz versus frequency are higher at lower frequencies. This is due to the flicker effect present in transistors. We mentioned earlier that flicker noise is predominant at low frequencies. Note also that the total noise voltage referred to the input versus source resistance increases as source resistance increases. This effect becomes pronounced above a source resistance of about 10 Kohm. Figure 10-15 lists some data for these graphs as they apply to a 741 op amp. This data must be stated for a particular temperature and supply voltage.

NOISE TESTS MADE ON THE 741 OPERATIONAL AMPLIFIER CONDITIONS: Vs = ±15V TA = 25°C		
FREQUENCY —HZ	MEAN SQUARE NOISE VOLTAGE —V^2/HZ	MEAN SQUARE NOISE CURRENT—A^2/HZ
10	5×10^{-15}	5×10^{-23}
100	1×10^{-15}	5×10^{-24}
1000	5×10^{-16}	8×10^{-25}
10,000	4.2×10^{-16}	3×10^{-25}
SOURCE RESISTANCE	TOTAL NOISE REFERRED TO INPUT μV RMS	
	BW 10HZ to 1 KHZ	BW 10HZ TO 10 KHZ	BW 10HZ TO 100 KHZ
100	0.8	2.0	6.0
1000	1.0	2.2	6.1
10,000	1.4	2.5	7.0
100,000	5.0	7.5	17.0

Fig. 10-15. Data for graphs in Fig. 10-14.

METHODS OF NOISE REDUCTION
IN CIRCUITS USING OPERATIONAL AMPLIFIERS

The following are guidelines in designing low noise operational amplifier circuits.

• If possible, reduce the size of the feedback resistor R2. Since noise currents from the operational amplifier flow through this resistor, a small value of resistance will result in small values of noise voltage.

• Use a bias compensating resistor if possible. This not only reduces the offset current problem in many operational amplifiers but also reduces noise currents since they are proportional to the offset currents. Refer to Fig. 10-16 to understand how offset currents are lowered. In Fig. 10-16A we see that there are unequal resistive paths for the currents into the inverting and noninverting input terminals. To correct this each input should see the same resistance to ground. The inverting input looks back into the parallel combination of R1 and R2. Therefore we should connect a compensating resistor R equal to R1R2/(R1 + R2) in series, the noninverting terminal and ground. Figure 10-16B shows how this resistor should be connected.

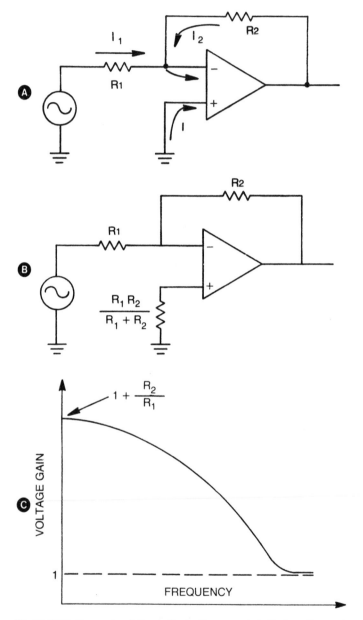

Fig. 10-16(A). Unequal resistive paths for the currents in the inverting and non-inverting input terminals. (B) Connection of compensating resistor. (C) As frequency becomes larger, noise gain falls off.

• Place a small capacitor across resistor R2 to reduce the noise gain at high frequencies. The noise gain then becomes that given by Equation 10-43.

Equation 10-43

$$A_n = 1 + Z_f/R_1$$

where Z_f is the parallel combination of a capacitor C in parallel with R2.

The quantity Z_f is R2/1 + jΩCR2. Placing this in Equation 10-43 we obtain Equation 10-44.

Equation 10-44

$$A_n = 1 + \frac{R_2}{R_1(1 + j\omega CR_2)}$$

We can see as W becomes larger, the noise gain reduces. This can be shown by Fig. 10-16C. At low frequencies the gain approaches 1 + R2/R1. However, at high frequencies the noise gain approaches unity. This helps reduce the overall noise voltage at the output.

• Never try to lower gain by placing a capacitor across R1. This only makes the noise gain increase with increasing frequency. The noise gain under this condition becomes that of Equation 10-45.

Equation 10-45

$$A_n = 1 + R_2(1 + j\omega CR_1)/R_1$$

This equation is plotted in Fig. 10-17. We can see that at low frequencies the gain is $1 + R_2/R_1$. However, as frequency ω continues to increase, so does the noise gain. This only makes the output noise voltage greater.

• Keep R_1 below or equal to about 10 kΩ.

Figure 10-18 illustrated an amplifier designed for low noise. The voltage gain is minus five. The largest noise gain that could ever occur is six, and a compensating resistor has been used.

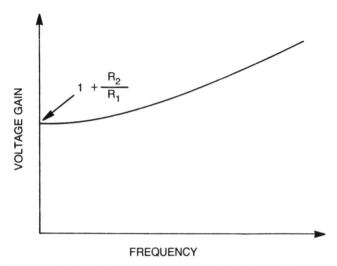

Fig. 10-17. Result of placing a capacitor across R1 in the amplifier.

Example 10-11

In Fig. 10-19 we see a simple inverting amplifier with a voltage gain of minus eight. Compute the compensating resistor to use in between noninverting input and ground and the noise gain at low frequencies.

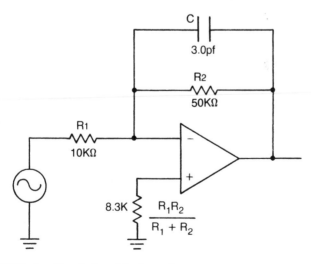

Fig. 10-18. Amplifier designed for low noise.

278

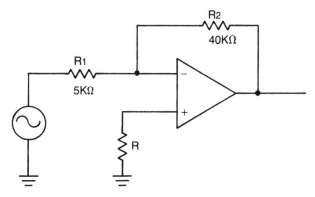

Fig. 10-19. Circuit for Example 10-11.

$$R = \frac{R1 \; R2}{R1 + R2}$$

$$= \frac{(5 \text{ k}) \, (40 \text{ k})}{5 \text{ k} + 40 \text{ k}}$$

$$= 5 \text{ k} \, (40)/45 = 4450\Omega$$

$$A_{nMAX} = 1 + R_2/R_1$$

$$= 1 + 40 \text{ k}\Omega/5 \text{ k}\Omega$$

$$= 1 + 8$$

$$= 9.$$

There are integrated circuit operational amplifiers now available which have field effect input transistors. The size of the offset currents may be extremely small such as a picoampere with a few having offset currents of a few femtoamperes. (One femtoampere is 10^{-15} amperes.) The noise currents generated by these FETs are therefore very small.

COMPLETE NOISE CALCULATION FOR THE CLASSIC NONINVERTING AMPLIFIER

In Fig. 10-20, we see the typical noninverting amplifier using an operational amplifier. In Fig. 10-21, the equivalent noise circuit is given. Resistors R1, R2, and R produce thermal noise. The noise produced by R1 can be shown as a voltage generator in series with

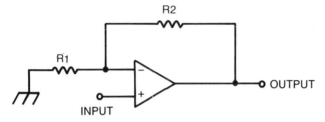

Fig. 10-20. Typical noninverting amplifier.

R1. This thermal noise is amplified by the gain of the inverting section and becomes that given by Equation 10-46.

Equation 10-46

$$(E_{n1}) \left(-\frac{R_2}{R_1} \right)$$

Equation 10-47

$$e_{h1} = (\sqrt{4\ kTBR1}) \left(-\frac{R_2}{R_1} \right)$$

The contribution due to R2 is simply that given by Equation 10-47.

Equation 10-48

$$e_{n2} = \sqrt{4kTBR_2}$$

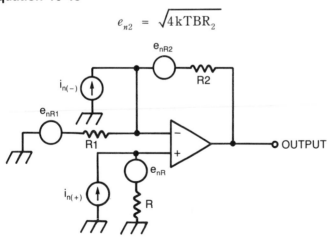

Fig. 10-21. Equivalent noise circuit for noninverting amplifier.

Notice that this is not multiplied by any gain. The reason for this is that resistor R_2 itself exists between the output and inverting input terminal.

Thermal noise due to resistor R is amplified by the noise gain or that given by Equation 10-49.

Equation 10-49

$$e_{n3} = \sqrt{4kTRB} \left(\frac{R_2}{R_1} + 1 \right)$$

Noise currents associated with the inverting and noninverting biasing currents can be given by noise current generators between the respective input terminals as shown.

Noise current $i_{n(-)}$ flows through R_2 producing a noise voltage as given by Equation 10-50.

Equation 10-50

$$E_{n4} = i_{n(-)} R_2$$

Noise current $i_{n(+)}$ flows through R producing a voltage drop. This drop is multiplied by the noise gain. This is given by Equation 10-51.

Equation 10-51

$$e_{ns} = i_{n(+)} R \left(\frac{R_2}{R_1} + 1 \right)$$

The amplifier noise voltage E_n can be shown in series with the source as shown in Fig. 10-22. This voltage becomes multiplied by the noise gain and becomes that given by Equation 10-52.

Equation 10-52

$$e_{n6} = E_n \left(\frac{R_2}{R_1} + 1 \right)$$

We can now include all of these voltages in a formula. We must remember that we are dealing with noise voltages that are uncorrelated. Therefore, we take the square of these sources, sum them, and find the square root.

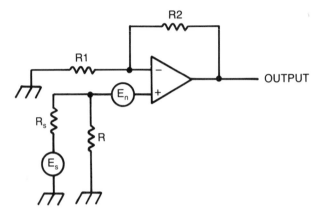

Fig. 10-22. Drawing showing location of E_N.

$$e_{nT} = \sqrt{e_{n1}^2 + e_{n2}^2 + e_{n3}^2 + e_{n4}^2 + e_{n5}^2 + e_{n6}^2}$$

Example 10-12

A certain op amp is used in the circuit shown in Fig. 10-20. The component values are as follows:

$$R1 = 10\ k\Omega$$
$$R2 = 100\ k\Omega$$
$$R = 9.1\ k\Omega$$

The bandwidth is 1 kHz. The lower frequency limit is zero hertz. The op amp noise currents are

$$i_{n(+)} = 50\ pA\ rms$$
$$i_{n(-)} = 50\ pA\ rms$$
$$E_n = 3\ \mu V\ rms$$

Compute all noise contributions and the total noise voltage. First of all, the rms noise contributions due to R1, R2, and R can be determined by the formula:

$$E_n = \sqrt{4\ k\ T\ R\ B}$$

For R1 we have:

$$E_{n1} = \sqrt{4 \times 1.38 \times 10^{-23} \times 293 \times 10^4 \times 10^3}$$

$$= 4.02 \times 10^{-7} \text{ volts}$$

$$= 0.402 \ \mu V$$

After amplification we have

$$e_{n1} = E_{n1} \left(-\frac{R_2}{R_1} \right)$$

$$= 0.402 \left(-\frac{100 \text{ k}}{10 \text{ k}} \right) \mu V$$

$$= 4.02 \ \mu V$$

For R2 we have:

$$e_{n2} = \sqrt{4 \times 1.38 \times 10^{-23} \times 293 \times 10^5 \times 10^3}$$

$$= \sqrt{1.617 \times 10^{-12}}$$

$$= 1.27 \ \mu V$$

For R we have:

$$e_{n3} = \sqrt{4 \times 1.38 \times 10^{-23} \times 293 \times 9.1 \times 10^3 \times 10^3}$$

$$= \sqrt{1.47 \times 10^{-13}}$$

$$= 3.83 \times 10^{-7}$$

$$= 0.383 \ \mu V$$

After gain,

$$e_{n3} = E_{n3} \left(1 + \frac{R2}{R1} \right)$$

$$= 0.383 \ \mu V \left(1 + \frac{100 \text{ k}}{10 \text{ k}} \right)$$

$$= 4.213 \ \mu V$$

$$e_{n4} = i_{n(-)} \, R2$$

$$= 50 \times 10^{-12} \, (10^5)$$

$$= 50 \times 10^{-7}$$

$$= 5 \ \mu V$$

$$e_{n5} = i_{n(+)} \, R \bullet \left(\frac{R_2}{R_1} + 1 \right)$$

$$= 50 \times 10^{-12} \times 9.1 \times 10^3 \, (11)$$

$$= \mu V$$

$$e_{n6} = E_n \left(\frac{R_2}{R_1} + 1 \right)$$

$$= 3\mu V \, (11)$$

$$= 33\mu V$$

$$e_{nT} = \sqrt{e_{n1}{}^2 + e_{n2}{}^2 + e_{n3}{}^2 + e_{n4}{}^2 + e_{n5}{}^2 + e_{n6}{}^2}$$

$$= \sqrt{(4.02)^2 + (1.27)^2 + (4.2)^2 + (5)^2 + (5)^2 + (33)^2} \ \mu V$$

$$= \sqrt{16.1 + 1.61 + 17.6 + 25 + 25 + 1089}$$

$$\cong \sqrt{1189}$$

$$= 34.5 \ \mu V$$

When designing amplifiers where extremely small noise is required, it is always advisable to investigate the possibility of using operational amplifiers which use field effect transistors.

Noise In Digital Integrated Circuits

Just like any circuit composed of transistors, diodes and resistors, the digital integrated circuit generates a certain amount of noise. In these circuits, however, we are working with pulses instead of analog type signals. Perhaps the most serious problem that develops regarding noise is that due to stray voltages picked up by the various wires connected to the digital integrated circuit.

Because of loading and temperature variations in logic circuits as well as component variations, the low and high voltage logic levels can vary over a large range of voltages. For example, suppose we have a NOR gate as shown in Fig. 10-23. A seven volt supply voltage is used in this circuit. The maximum high logic level is seven volts while the minimum high logic level might be three volts. Usually we will find several gates connected together. It is a necessary design requirement that the minimum high level output will be large enough to drive another NAND gate. Should this voltage be too small, it will not be capable of driving the next gate to the one condition. We will call this voltage V_{HOM} meaning high logic level output minimum value. The driven logic circuit requires a certain minimum input high level voltage to affect turn on. This value will be called V_{HIM}, meaning high logic level input minimum value required for turning on a gate. Assume that for the gate we

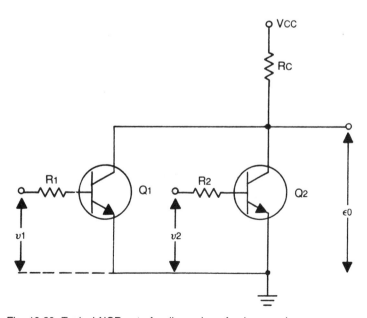

Fig. 10-23. Typical NOR gate for discussion of noise margin.

are discussing the value of V_{HIM} is two volts. The gate we are discussing then requires a minimum of two volts to turn on its input circuit. The output of the NOR gate can drive another gate since the minimum value of output is three volts or one volt larger than V_{HIM}. This is the smallest difference that can ever occur. The largest difference possible would be the maximum value for the high level output or seven volts minus two volts or five volts.

The difference between V_{HOI} and V_{HIM} is called the noise margin and is given by Equation 10-53.

Equation 10-53

$$N.M. = V_{HOM} - V_{HIM}$$

A noise margin of one volt means that a noise signal of at least one volt would be required to cause the output voltage to drop below the minimum high level output voltage required. The higher the noise level, the more immune to noise the circuit happens to be. It is more likely for small noise voltages to exist than large noise voltages. When a circuit has a low noise margin, we can say it also has a low noise immunity. However, if a circuit has a large noise margin, it would have a high noise margin. If the noise level had been two volts, the more immune to noise the logic gate would have been.

Types of Noise Which May Effect A Digital Logic Circuit

The noise which may cause problems in logic circuit operation may be caused by situations like unstable supply voltages, transient radiated from other circuits or static phenomena and inductive effects.

Just as there was a high level noise margin, there is also a low level margin. This represents a measure of the amount of noise which can be superimposed on a low voltage logic level and still cause no change of logic state.

In Fig. 10-24A, we see some typical noise margins for discrete logic circuits. In Fig. 10-24B are some typical noise margins for digital integrated circuits.

One major cause of noise which may switch the logic circuits randomly is due to transients in electronic equipment. Transients may be caused by a variety of disturbances such as electric motors, welding equipment, and static disturbances. It is well known

Noise Margins for Discrete Digital
Circuits

High Voltage Noise Margin

RTL	Volts
DTL	Volts
TTL	Volts

(A)

Comparison of Noise Generation and Noise Immunity
for Four Types of Digital Integrated
Circuit Families—Relative

	TTL	ECL	CMOS
Noise Generation	High	Low	Medium
Noise Immunity	Good	Good	Excellent

(B)

Fig. 10-24(A). Typical noise margins for discrete logic circuits. (B) Typical noise margins for digital integrated circuits.

that a computer system's memory can under certain circumstances be "knocked out" by a power line failure due to a lightning induced transient. In a digital system, the most severe types of transients occur from the rapid switching of current and voltage. Current switching can be broken into three categories: spikes from power supplies, ground noise and induced noise. Voltage switching may be considered capacitive coupled noise.

Power Supply Spiking

Suppose a large amount of current is drawn suddenly from a power supply terminal. Since the power supply has a certain amount of internal resistance, an internal voltage drop will occur across the internal resistance and a drop in power supply terminal voltage will occur. Since the power supply is possibly serving several circuits, there will be a spike on the supply voltage of every circuit. This spike may cause undesirable triggering of such circuits as flip flops and may also cause ringing to appear. To reduce this effect, the currents drawn from the power supply during switching should be kept small. However, this is not always possible and

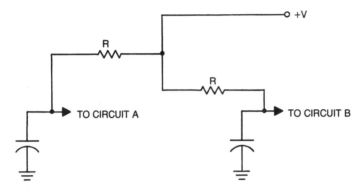

Fig. 10-25. Power supply spiking noise.

some more elaborate method is required. One way is to use a decoupling network between the power supply and the various circuits. Should any sudden narrow pulses appear on the power supply output, they will be prevented from entering other circuitry by this low pass network. This circuit is shown in Fig. 10-25.

Induced Noise

Whenever wires are near each other, there is the possibility of an induced noise signal from one wire to another. This occurs because when the current changes in wire X, there will be a change in magnetic flux about wire Y. However, wires X and Y are near each other and the change of flux about wire X will link wire Y. This change in flux in wire Y will result in an induced voltage in wire Y. The best way to lower the effects of this linkage flux is to twist the two wires together. This tends to cancel the effects of this flux. Figure 10-26A illustrates how the induced voltage develops while Fig. 10-26B shows how we remedy this effect.

Ground Noise

In an electronic system, it is only natural for all of the circuits to have the same ground. As current changes through one of the circuits, a voltage drop may occur in the ground return wire due to the product of wire inductance and the rate of change of current in respect to time. This is given by Equation 10-54.

Equation 10-54

$$V_m = L_w \, d_i/d_t$$

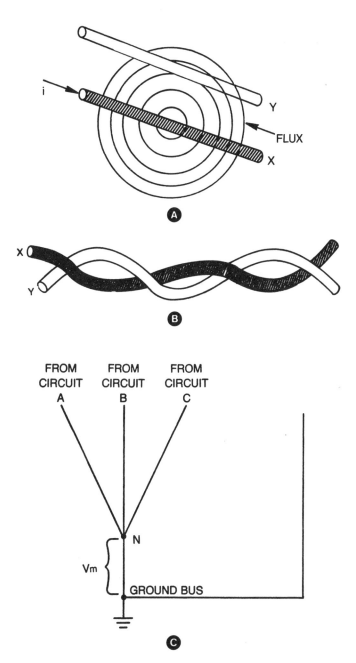

Fig. 10-26(A). Development of induced voltage. (B) How to remedy the effect of induced voltage. (C) Development of ground noise.

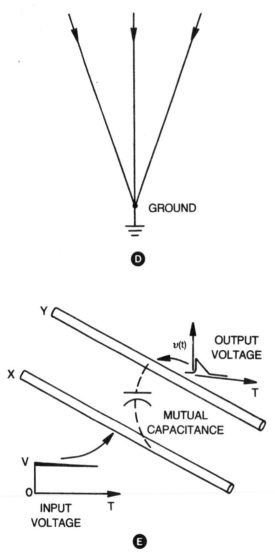

Fig. 10-26. (D) How to remedy ground noise. (E) Conditions necessary for the development of capacitive coupled noise.

where V_w is the voltage induced in the wire, L_w is the inductance of the wire, d_i/d_t is the rate of change of current in respect to time.

In Fig. 10-26C we can see what happens in a circuit due to the ground noise. Suppose circuit A causes a ground noise voltage to develop at point N. This pulse is also felt at circuit B and C. If we were to tie circuits A, B, and C individually to ground, very little induced noise will be developed and shared between each circuit. This is shown in Fig. 10-26D.

Capacitive Coupled Noise

When two conductors are placed in close proximity there is a capacitance between the two wires. Should there be a change in voltage on one of the wires, there will be a voltage developed on wire Y. The larger the capacitance between wire Y to ground, the larger this voltage becomes. In many cases this capacitance is small and very little problems occur due to the capacitive coupled noise. To reduce the effects of this noise do the following:

- Circuits that may be triggered by false inputs should have small input impedances.
- Use twisted pairs of wires.
- Use metallic shielding between the wires.
- Try placing all wires near the chassis if possible.

PARD

The term *PARD* refers to both ripple and noise taken together as an undesirable disturbance at the output of a power supply. This may be given as a peak-to-peak value or rms value for a specified bandwidth. Pard stands for *periodic and random deviation.*

To understand PARD, let us investigate two basic power supplies and see if we can obtain some insight into how we could forecast this ripple and noise.

In Fig. 10-27A, we see the basic half-wave power supply. We know that as the input voltage goes through its positive alternation the diode conducts and a positive alternation of voltage occurs across the resistor R.

The frequency make-up of this signal can be derived as follows (refer to Fig. 10-27B). Since the rectifier acts like a switch, we can imagine that a switching function exists in the circuit. We will assume that when the positive alternation of input comes in, the signal is multiplied by unity. However, when the signal goes negative, it is multiplied by zero.

In Chapter 2, we derived the Fourier series for a signal of the form of our switching function. This was derived in Example 2-9. If we allow the switching function to have an amplitude of unity,

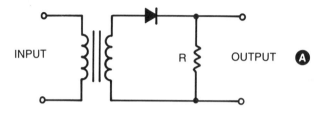

Fig. 10-27A. Basic half-wave rectifier.

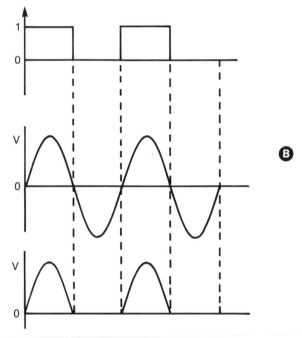

Fig. 10-27B. Switching function for half-wave rectifier.

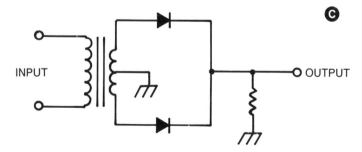

Fig. 10-27C. Basic full-wave rectifier.

then the Fourier make-up could be written as Equation 10-55.

Equation 10-55

$$S_{(t)} = \frac{1}{2} + \frac{2}{\pi} \cos \frac{2 \pi t}{T} - \frac{2}{3 \pi} \cos \frac{6 \pi t}{T}$$

$$+ \frac{2}{5 \pi} \cos \frac{10 \pi t}{T}$$

If we say that the input to the power supply is a cosine waveform, we can write the output signal as being equal to the input multiplied by the switching function. The result would be given by Equation 10-56.

Equation 10-56

$$e_o = e_i (S_{(t)})$$

$$= \left[V \cos \omega t \right] \left[\frac{1}{2} + \frac{2}{\pi} \cos \frac{2 \pi t}{T} - \frac{2}{3 \pi} \cos \frac{6 \pi t}{T} \right.$$

$$\left. + \frac{2}{5 \pi} \cos \frac{10 \pi t}{T} \cdots \right]$$

$$e_o = \frac{V}{2} \cos \omega t + \frac{2 V}{\pi} (\cos \omega t) (\cos \frac{2 \pi}{T} t)$$

$$- \frac{2 V}{3 \pi} (\cos \omega t) (\cos \frac{6 \pi}{T} t)$$

$$+ \frac{2 V}{5 \pi} (\cos \omega t) (\cos \frac{10 \pi}{T} t)$$

Now we know that $\frac{2 \pi}{T}$ is the fundamental frequency and $\frac{6 \pi}{T}$ would be the third harmonic. Therefore, we may rewrite the above as Equation 10-57.

Equation 10-57

$$e_o = \frac{V}{2} \cos \omega t + \frac{2 V}{\pi} (\cos \omega t) (\cos \omega t)$$

$$- \frac{2V}{3\pi} (\cos \omega t)(\cos 3\omega t)$$

$$+ \frac{2V}{5\pi} (\cos \omega t)(\cos 5\omega t)$$

This may be expanded as follows:

$$e_0 = \frac{V}{2} \cos \omega t + \frac{2V}{\pi} \left[\frac{\cos 2\omega t + \cos 0}{2} \right]$$

$$- \frac{2V}{3\pi} \left[\frac{\cos 4\omega t + \cos 2\omega t}{2} \right]$$

$$+ \frac{2V}{5\pi} \left[\frac{\cos 6\omega t + \cos 4\omega t}{2} \right]$$

$$e_0 = \frac{V}{2} \cos \omega t + \frac{V}{\pi} \cos 2\omega t + \frac{V}{\pi}$$

$$- \frac{V}{3\pi} \cos 4\omega t - \frac{V}{3\pi} \cos 2\omega t$$

$$+ \frac{V}{5\pi} \cos 6\omega t + \frac{V}{5\pi} \cos 4\omega t$$

We can combine these and obtain Equation 10-58.

Equation 10-58

$$e_0 = \frac{V}{\pi} + \frac{V}{2} \cos \omega t - \frac{2V}{3\pi} \cos 2\omega t$$

$$- \frac{2V}{15\pi} \cos 4\omega t - \frac{2V}{35\pi} \cos 6\omega t$$

$$- \frac{2V}{[n^2 - 1]\pi} \cos n\omega t$$

Where only even values of n

Suppose we are working with a power supply that operates on the typical frequency of 60 Hz. Then the Fourier series would be given by Equation 10-59.

Equation 10-59

$$e_o = \frac{V}{\pi} + \frac{V}{2} \cos 2\pi (60) t - \frac{2\,V}{3\,\pi} \cos 2\pi (120) t$$

$$- \frac{2\,V}{15\,\pi} \cos 2\pi (240) t$$

$$- \frac{2\,V}{35\,\pi} \cos 2\pi (360) t$$

In a similar manner, we could evaluate the frequency make-up of the output of a full-wave rectifier. In Fig. 10-27C, we see such a circuit. The switching function of such a network can be viewed as a function that is unity when the positive alternation comes in, while it becomes negative when the negative alternation comes in. The Fourier series was derived for this in Chapter 2. The series for unit amplitude could be given as Equation 10-60.

Equation 10-60

$$S_{(t)} = \frac{4}{\pi} \sin \frac{2\pi t}{T} + \frac{4}{3\pi} \sin \frac{6\pi t}{T}$$

$$+ \frac{4}{5\pi} \sin \frac{10\pi t}{T}$$

The output for such a circuit is also the product of the input voltage and the switching function. If the input is a sine wave of equation $V \sin wt$, we can write the output as Equation 10-61.

Equation 10-61

$$e_o = (e_{in}) (S_{(t)})$$

$$= (V \sin \omega t) \frac{4}{\pi} \sin \omega t + \frac{4}{3\pi} \sin 3\ \omega t$$

$$+ \frac{4}{5\pi} \sin 5 \ \omega t + \ --- \ \Bigg]$$

$$= \frac{4V}{\pi} \sin^2 \omega t + \frac{4V}{3\pi} (\sin 3 \ \omega t)(\sin \ \omega t)$$

$$+ \frac{4V}{5\pi} (\sin 5 \ \omega t)(\sin \ \omega t)$$

$$e_o = \frac{4V}{\pi} \Bigg[\frac{1}{2} - \frac{1}{2} \cos 2 \ \omega t \Bigg]$$

$$+ \frac{4V}{3\pi} \Bigg[\frac{1}{2} \Bigg] \Bigg[\cos 2 \ \omega t \ - \ \cos 4 \ \omega t \Bigg]$$

$$+ \frac{4 V}{5\pi} \Bigg[\frac{1}{2} \Bigg] \Bigg[\cos 4 \ \omega t \ - \ \cos 6 \ \omega t \Bigg]$$

$$+ \ . \ . \ .$$

This may be then restated as Equation 10-62.

Equation 10-62

$$\int e_o = \frac{2 V}{\pi} - \frac{2 V}{\pi} \cos 2 \ \omega t + \frac{2 V}{3\pi} \cos 2 \ \omega t$$

$$- \frac{2 V}{\pi} \cos 4 \ \omega t + \frac{2 V}{5\pi} \cos 4 \ \omega t$$

$$- \frac{2 V}{5 \pi} \cos 6 \ \omega t$$

Equation 10-62 then reduces to Equation 10-63.

Equation 10-63

$$e_o = \frac{2 V}{\pi} - \frac{4 V}{3\pi} \cos 2 \ \omega t - \frac{4 V}{15\pi} \cos 4 \ \omega t$$

$$- \frac{4 V}{35 \ \pi} \cos 6 \ \omega t \ . \ . \ .$$

For a 60 Hz input frequency, the output therefore becomes equal to Equation 10-64.

Equation 10-64

$$e_o = \frac{2\,V}{\pi} - \frac{4\,V}{3\pi} \cos 2\pi \, (120) \, t$$

$$- \frac{4\,V}{15\pi} \cos 2\pi \, (240) \, t$$

$$- \frac{4\,V}{35\pi} \cos 2\pi \, (360) \, t$$

It is obvious that if no filtering was present, the ripple content would be at maximum and therefore the PARD value would be at maximum.

A manufacturer may rate a power supply in regards to PARD as follows:

> PARD ·
> output ·   80 mV p – p (30 kHz to 10 MHz)

Should the noise be small compared to the ripple, then the PARD is mainly the ripple. PARD can be reduced by improving the amount of filtering.

Now that we have seen some of the quantities that determine PARD, we can easily see why the term stands for periodic and random deviation of the output from its average level. The measurement of PARD must be taken into account with external and environmental parameters remaining constant. The most important parameter that must remain constant is the load impedance. Variations in the load impedance result in disturbances such as dynamic regulation and transient response.

Suppression of Noise in Communication Systems

Noise produces the most problems in amplitude modulation communication systems since noises seem to disturb the amplitude more than the frequency of a radio signal. Frequency modulation circuits are not affected to any great extent due to amplitude limiter circuits. For citizens band equipment many steps may need to be taken to help eliminate or suppress noise. A large amount of noise

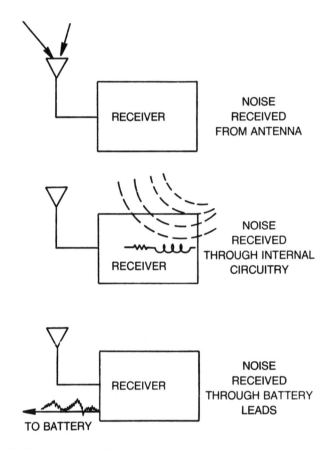

Fig. 10-28. Methods by which noise may enter a receiver.

is generated in automobiles where CB is quite popular. Likewise, mobile installations in airplanes and boats suffer from noise generated by the electrical and ignition systems.

Noise may enter a receiver through radiation reception in the internal circuitry, through the antenna and through the wires leading to the battery.

These are illustrated in Fig. 10-28. Now let us look at some of the most common types of noise that occur in a vehicular system.

Ignition System

The ignition system produces a noise which results in a popping sound that may be quite serious when the engine is running slow. The causes of this type of noise may be due to the normal ignition system operation. Whenever an electric spark occurs, there are many frequency components which can interfere with the reception of radio signals. Also, discharges take place between the engine and wiring, across a gap of a poor connection or around the casing of a spark plug which has grease on it.

Some of the remedies for ignition noise are listed below:

• Have a major tune-up of engine. Have new spark plugs installed and new ignition capacitor.

• Shield the ignition system.

• Install an ignition suppressor in series with the center high tension lead connected to the distributor. Refer to Fig. 10-29A. It is also possible to install suppression wire in the system as shown in Fig. 10-29B to alleviate the problem. Adding wire to the ignition circuit results in a lowering of ignition noise.

Generator Noise

The vehicles generator produces a noise which is a big problem at high frequencies. As the generator armature turns, arching occurs between the generator brushes and the commutator. The type of noise that you will notice in the loudspeaker is a whining sound. This whine varies with the speed of the engine. One way of testing to see if the noise is coming from the generator is to shut off the ignition. You should hear a decreasing whine as the generator comes to a halt. The alternator may cause similar effects in the system. The steps to reduce this noise are as follows:

1. Recondition the alternator or generator.

2. Install a 0.5 microfared coaxial type capacitor in the generator armature lead or in series with the alternator output. This is illustrated in Fig. 10-29C. The capacitor rating should be in the range of 200 amperes.

3. Check to see if alternator slip rings and generator brushes are clean and unworn.

4. Connect a noise trap as shown in Fig. 10-29D in series with the armature lead to the generator. This trap acts like a parallel tuned circuit which produces a high impedance to those noise signals whose frequencies lay within its bandwidth. Therefore a large percentage of the noise components produce voltage drops across the tuned circuit. There is very little noise to get into the electrical system wires. Figure 10-29E illustrates the electrical make-up of this filter.

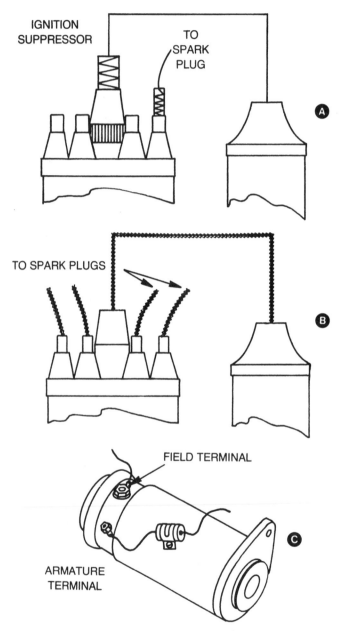

Fig. 10-29(A). Installation of an ignition suppressor in series with the high tension lead connected to the distributor. (B) Using suppression wire instead of an ignition suppressor. (C) A 0.5 microfarad capacitor may be connected in the generator armature lead.

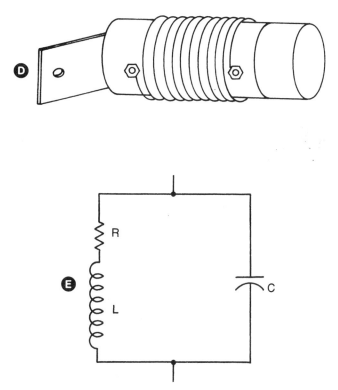

(D) A noise trap may be connected in series with the armature lead to the generator. (E) Electrical makeup of a noise trap.

Voltage Regulator

The voltage regulator produces a hash type of sound which occurs as the relay points close and open. A reduction in the noise occurs by installing bypass capacitors in series with the battery and armature leads of the regulator. See Fig. 10-30 for an illustration. Figure 10-31 lists a few more of the noise sounds which occur in the loudspeaker and their likely cause.

All of the previous information has been presented as an introductory topic to noise in a mobile system. For further information consult many excellent books available on the electrical system of automobiles.

COMMUNICATION SYSTEM USING A FIBEROPTIC LINK

We already know that a considerable amount of noise may be

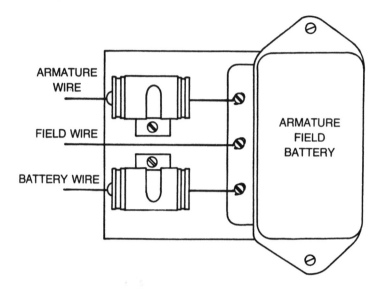

Fig. 10-30. Reduction of hash in a voltage regulator.

brought into a communication system's receiver due to noise in the channel. If we could make the channel unaffected by this noise, we would have the answer to this problem. A communication system which uses a fiber optic link between transmitter and receiver will help greatly in this situation.

What Are Fiberoptics?

Fiberoptics may be defined as the technique of conveying light or images through a particular configuration of glass or plastic fibers.

We may divide the subject into three subheadings: incoherent fiber optics, which will transmit light as a pipe will transmit water—but not an image; coherent fiber optics, which can transmit light in the form of an image through well aligned optical fibers; and

NOISE HEARD IN LOUDSPEAKER	SOURCE
CRACKS AND HISSES	INSTRUMENT GAUGES
CLICKS	GAUGE SENDERS
RANDOM POPPING	STATIC FROM WHEELS
INTENSE REGULAR POPPING	TIRE STATIC

Fig. 10-31. Other causes of noise which occur in a mobile system.

specialty fiber optics, which combines incoherent and coherent aspects.

Fiberoptics has the potential of many significant advantages compared to metallic conductors because of low loss transmission achievements in the optical fiber.

In a communication system, one of the major problems that exists is that of electrical noise. Although the use of a optical cable between transmitter and receiver will not help reduce internal circuit noise, it serves as another means of conveying information. Thus we are not bothered by such things as atmospheric noise, etc.

Some of the advantages of fiber optics are long range transmission without repeaters, crosstalk and ground loop immunity, small size and weight and high degree of intercept security and dielectric isolation.

The desirable features of optical fiber waveguides have stimulated efforts in supporting technologies such as fiber cabling, couplers and long life solid state sources and receivers.

The fiber optical waveguide has become the leading contender as the transmission medium for a variety of systems ranging in length form 10 to 10^4 meters.

Light Transmission in an Optical Fiber

Shown in Fig. 10-32 is an optical fiber. The cladding is a sheathing or covering, usually of glass, fused to the core of the fiber. The core is the center dielectric whose index of refraction is larger than that of its surroundings medium, usually a clodding.

Light may enter the fiber with any angle equal to or less than 2Θ ac. The angle which we call 2Θ ac is known as the total acceptance angle. One half this angle is Θ ac. The side of the angle is called the "numerical aperture." Refer to Fig. 10-33.

This is a number that expresses the light gathering ability of the fiber. The "degree of openness," "light-gathering ability," and acceptance cone are all terms describing the characteristics. The numerical aperture is namely:

$$N.A. = \sqrt{(n_1^2 - n_2^2)}$$

where n1 = refractive index of core
n2 = refractive index of cladding
n = refractive index of outside

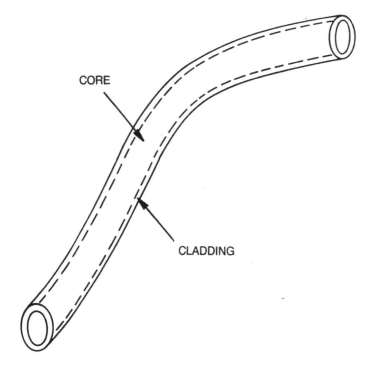

Fig. 10-32. Typical fiberoptic cable showing core and cladding.

Example 10-12

For a particular fiber, the index of refraction of the core is 1.6 and the cladding refraction index is 1.4. Find the total acceptance angle of the fiber, assuming the outside refractive index is unity.

$$\sin \Theta ac = \sqrt{(n_1^2 - n_2^2)}/n$$

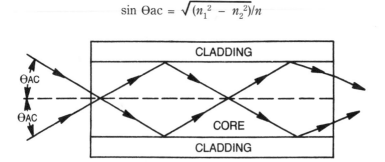

Fig. 10-33. Illustration showing the acceptance angle.

$$= \sqrt{(1.6)^2 - (1.4)^2/1} = \sqrt{2.56 - 1.96}$$

$$= \sqrt{.6} = .77$$

$$\Theta\,ac = 50 \text{ degrees}$$

Total acceptance angle = $2\Theta ac$ = 100 degrees

As light is transmitted through the cable, there is little chance for outside noise to affect the transmission. It is almost impossible for outside light to penetrate the fiber cable. However, if it would, this light would be considered an optical noise signal. Due to the high immunity of the cable to outside light sources, fiberoptic cables are being used in the development of various types of communication systems.

Problems of the Fiberoptic Link

Whenever light is transmitted through a fiber, several things occur which tend to alter the appearance of the light pulse at the output of the fiber. First, there is attenuation. All cables have a certain amount of attenuation. This attenuation varies with the various wavelengths in the light pulse being transmitted. Figure 10-34

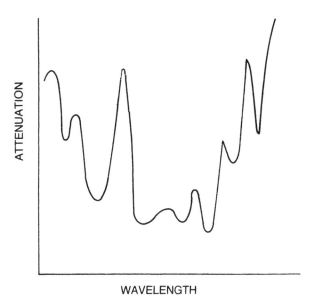

WAVELENGTH

Fig. 10-34. Attenuation versus wavelength.

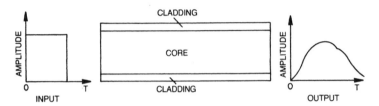

Fig. 10-35. Effect of dispersion on a light pulse in a fiberoptic cable.

shows a plot of attenuation versus wavelength. Some cables may have very large attenuations such as 5000 decibels per kilometer for some wavelength while another cable may have a very low attenuation such as five decibels for the same wavelength.

As light travels through the fiber, there are many different paths for it to take. When we regroup these light rays at the output, the resulting pulse looks distorted in respect to the input pulse. The pulse width seems to be widened. The parameter which causes this is called modal dispersion. Modal dispersion is a number which describes how much a light pulse will be increased in width as it passes through the cable. The unit of modal dispersion is nonoseconds per kilometer. Figure 10-35 shows a light pulse entering a fiber and its resultant at the output notice that the width of the output pulse is greater due to dispersion.

Example 10-13

A light pulse of 40 nanoseconds is applied to a cable. If the cable is 0.1 km in length, how wide will the output light pulse be if the modal dispersion is 20 nsec/km?

Use the following formula to calculate the width of the output pulse.

$$\omega_{out} = \omega_{IN} + (Gg)(L)$$

where ω_{out} is the outside pulse width,
ω_{IN} is the input pulse width,
Gg is the modal dispersion,
L is the fiber optic cable length.
Placing the various quantities in this formula we have:

$$\omega_{out} = 40 \text{ nsec} + (20 \text{ nsec})/\text{Km} (.1\text{Km})$$
$$= 42 \text{ nsec}.$$

Another problem is material dispersion caused by various frequency dependent properties in the fiber. This effect is not as serious as that due to modal dispersion. The result of material dispersion is also an increase in pulse width at the output.

Although the problems of attenuation and dispersion occur with the transmission of light through a fiber optic cable, they can be predicted. However, if we had used the atmosphere as the communication medium, we would be perplexed with such problems as precipitation and dust particles. All of these cause random variations in the light information being transmitted. These effects cannot be predicted and they will enhance the noise level at the receiver. Also, outside light sources will affect the transmission of a modulated light beam for the optical communication that does not use a fiber optic link.

Examples of Optical Communications Systems

In Fig. 10-36 is shown an optical communication system using the atmosphere as the communication medium. The audio signal is applied to an amplifier circuit which feeds a light emitting diode. As the forward current of the diode is varied, the optical output varies. This power is transmitted towards the receiver which uses a photodiode or phototransistor for the detector. Both of these devices generate a certain amount of noise due to such things as a system not using an fiber optic link. The range of the system is inversely proportional to the square root of the detector's sensitivity. This sensitivity is the smallest power level which can be received in the presence of noise. Since the range is proportional to the condition of the atmosphere, such things as precipitation and dust greatly cause the transmission ability to be reduced. If we connect a fiber optic cable between the transmitter and receiver as shown in Fig. 10-37, we will notice a decrease in the noise level and, in many instances, an improvement in signal strength coming from the receiver.

In Fig. 10-38 is shown a basic optical communication system. The light emitting diode transmits light at the detector with divergence angle Θ. The maximum range is by Equation 10-48.

Equation 10-65

$$R_{MAX} = \sqrt{P_o A_r \; \copyright \; ka/P_{th} (\Theta)^2}$$

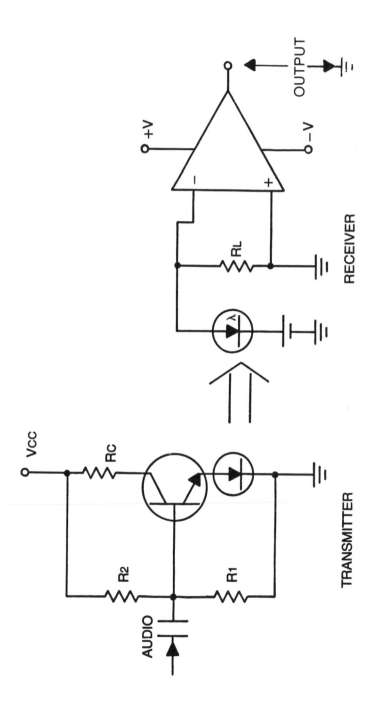

Fig. 10-36. Optical communication system using the atmosphere as the communication medium.

FIBER OPTIC CABLE

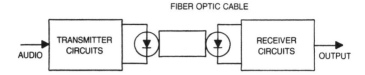

Fig. 10-37. Optical communication system using a fiberoptic cable.

where P_o is the LED output power,

P_{th} is the sensitivity of the detector,

ka is the atmosphere constant,

Θ is the divergence angle,

Ar is the detector area.

Example 10-14

Compute the range of an optical communication which has the following area:

$$P_o = 1 \text{ Watt}$$
$$A_r = 10^{-3} \text{ meters}^2$$
$$ka = 0.64$$
$$\Theta = 0.1 \text{ radians}$$
$$P_{th} = 10^{-5} \text{ watts}$$

$$R_{MAX} = \sqrt{\frac{(1)\ (10^{-3})\ (0.64)}{10^{-5}\ (1 \times 10^{-1})^2}}$$

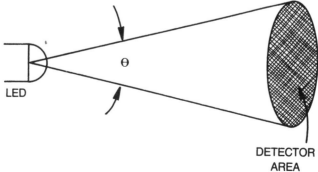

Fig. 10-38. Basic optical communication system.

$$= 0.64 \times 10^4$$

$$= .8 \times 10^2$$

$$= 80 \text{ meters.}$$

NOISE REDUCTION SYSTEM FOR RECORDING SYSTEMS

One important system which was able to reduce audible noise heard on cassettes was invented by Dr. Ray Dolby. The system is referred to as the Dolby noise reduction system. The system provides and extra 10 dB of signal to noise ratio to reduce hiss, hum and distortion. Quite frequently the type of noise heard on cassettes is a hiss which is made of high-frequency noise. The system changes high frequency component amplitudes by an amount dependent on the speech or music volume at any given time. When the signal is loud, noise becomes mashed and the Dolby system does not boost any of the high frequencies. When the music or speech are soft, the Dolby system gives a boost to the high frequency components which are being recorded. As the music or speech becomes softer, the system gives more boost to the higher frequencies.

During playback, high frequencies have their amplitudes attenuated during soft volumes. During loud volumes the frequency components are not attenuated. The result is that when the high frequencies are reduced in amplitude, so are the noise components. The overall effect is that the speech or music seems to be unaffected because noise has been reduced. The signal components which are reduced during the playback were increased during recording. Figure 10-39 shows the system operation for recording and playback as a simplified block diagram of the system.

In Fig. 10-39 we see that the input signal is split by filters into four frequency bands. They are a low pass with cutoff at 80 Hz, bandpass with cutoffs of 80 Hz and 3 kHz respectively, high pass with cutoff at 3.3 kHz and a high pass with cutoff at 9 kHz.

Each frequency band is separately compressed, so that low signals around -40 dB are increased by 10 dB while signals of 0 to -30 dB are not affected.

On playback the signal is returned to normal by use of an expansion network. Noise which is similar in frequency to the louder components of the signal will be reduced in level.

Noise in Optoelectronic Devices

When dealing with optoelectronic devices, the noise may be

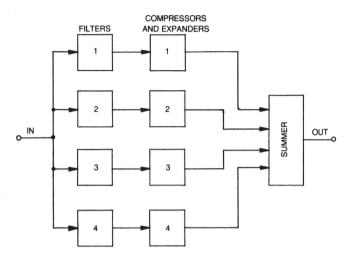

Fig. 10-39. Block diagram for the Dolby system.

fluctuations in the output voltage or current. The amount of noise is directly related to the bandwidth of the circuit the device is being used in. Since detectors have a certain "capture area", the noise is related to this area. As a rule of thumb, it can be said that the noise produced is proportional to the square root of the detector area.

It is quite simple to define a signal-to-noise ratio for a optoelectronic device. This of course occurs at the output. However, it is extremely difficult to define a noise figure. This is because the input power to the device is optical and the noise power is also optical.

The random variation in the optical input power could be defined as optical input noise. This variation is caused by dust, precipitation, and sintillation. For example, the noise figure for an optoelectronic device could be defined as Equation 10-66.

Equation 10-66

$$F = \frac{\dfrac{S_{io}}{N_{io}}}{\dfrac{S_o}{N_o}}$$

where

S_{io}	is the optical input signal
N_{io}	is the optical input noise power
S_o	is the signal output power
N_o	is the noise output power

The signal output power could be given by the Equation 10-67.

Equation 10-67

$$S_o = I^2_{signal} R$$

where I signal is the rms current in the output due to the optical input signal power.

The noise output power could be given by the Equation 10-68.

Equation 10-68

$$N_o = I_{noise}^2 R$$

where I noise is the rms noise current in the output due to the optical input noise power.

Since it is a bit difficult to define a noise figure for such a device, the manufacturer may specify an equivalent power for noise. This is defined by Equation 10-69.

Equation 10-69

$$\frac{\text{Noise Equivalent Power}}{} = \frac{\text{Noise Current}}{\text{Responsivity}}$$

The noise current is given in amps per square-root of hertz, whereas the responsivity will be in amps/watt. The noise equivalent power will have a unit of watt per square-root of hertz. This term is frequently called "NEP."

NOISE AND DISTORTION

At this time it is only right to discuss the difference between noise and distortion. Noise usually is thought of as undesirable electrical signals introduced by equipment or natural disturbances. Distortion, is the result of the transmission characteristics of a medium itself.

Examples of Distortion

Suppose a sinewave of the form $A \sin wt$ is applied to a circuit with the characteristic shown below:

$$e_o = k\, e_{in}^2$$

The output will become:

$$e_o = k\,(A \sin^2 \omega t)$$

$$= k\left[A^2\right]\left[\frac{1}{2} - \frac{1}{2}\cos 2\,\omega t\right]$$

$$= \frac{k\,A^2}{2} - \frac{k\,A^2}{2}\cos 2\,\omega t$$

We started with a simple sinewave, but ended up with two terms. The term $KA^2/2$ is called a rectification product or dc term. The second term $- KA^2/2 \cos 2wt$ is the second harmonic of the sinewave. This means that the waveform has a frequency that is twice the original frequency. We also noted that the waveform is a cosine instead of a sine. If the input had been a cosine such as $A \cos wt$, the output would be as follows:

$$e_o = k\, e_{in}^2$$

$$= k\,(A^2 \cos^2 \omega t)$$

$$= \frac{k\,A^2}{2} + \frac{k\,A^2}{2}\cos 2\,\omega t$$

Notice that we still obtain a cosinewave and a dc term. Now, if the output characteristic would be:

$$e_{out} = k\, e^3_{in}$$

$$e_{out} = k\, e_{in} \times e_{in}^2$$

$$= (k\,A \cos \omega t)\,(A^2 \cos^2 \omega t)$$

$$= \left(A \cos \omega t \right) \left(\frac{kA^2}{2} + \frac{kA^2}{2} \right) \cos 2 \, \omega t$$

$$= \frac{k \, A^3}{2} \cos \omega t$$

$$+ \left(k \, A \cos \omega t \right) \left(\frac{A^2}{2} \cos 2 \, \omega t \right)$$

$$= \frac{k \, A^3}{2} \cos \omega t$$

$$+ \left(\frac{k \, A^3}{4} \cos \omega t + \frac{k \, A^3}{4} \cos 3 \, \omega t \right)$$

$$= \left[\frac{k \, A^3}{2} + \frac{k \, A^3}{4} \right] \cos \omega t$$

$$+ \frac{k^2 \, A^3}{4} \cos 3 \, \omega t$$

We see that we have now ended up with the original frequency w as well as the third harmonic. Therefore with the characteristic $KE^2{}_{in}$, we get only a dc term and the second harmonic, but $E_{in}{}^3$ gives us the original frequency as well as the third harmonic.

Let us take one more situation. Suppose we have a characteristic that gives $kE^4{}_{in}$.
The output will then be of the form

$$K \, (A \cos \omega t)^4$$

This gives us:

$$k \, (A^2 \cos^2 \omega t) \, (A^2 \cos^2 wt)$$

$$k \left(\frac{A^2}{2} + \frac{A^2}{2} \cos 2 \, \omega t \right) \left(\frac{A^2}{2} + \frac{A^2}{2} \cos 2 \, \omega t \right)$$

$$= k \left(\frac{A^4}{4} + \frac{A^4}{2} \cos 2 \, \omega t + \frac{A^4}{4} \cos^2 2 \, \omega t \right)$$

$$= k \left(\frac{A^4}{4} + \frac{A^4}{2} \cos 2 \omega t + \frac{A^4}{8} + \frac{A^4}{8} \cos 4 \omega t \right)$$

We now notice that the second harmonic, fourth harmonic, and a dc term occur. Therefore, with the additional waveforms being generated, it is only right to admit that the shape of the output waveform will be different from that of the input waveform.

THE DIFFERENTIAL AMPLIFIER

One very valuable circuit that is used in various circuits to cancel the effect of noise is known as the differential amplifier. Such an amplifier amplifies the difference between two signals. Such a circuit has the advantage of the canceling noise that is common to the two input signals. In Fig. 10-40, we see the basic block diagram. Inputs E_1 and E_2 are returned to ground at the same point, however, the wire which returns to ground has some signal developed in it due to some property such as thermal noise or noise induced by another wire in proximity to the ground wire. The circuit could be redrawn as in Fig. 10-41. Since the signal on the one input is $e_1 + e_n$, and on the other input is $e_2 + e_n$. The output is given by Equation 10-70.

Equation 10-70

$$e_0 = A \, [e_1' - e_2']$$

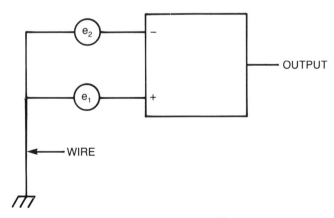

Fig. 10-40. Block diagram of a differential amplifier.

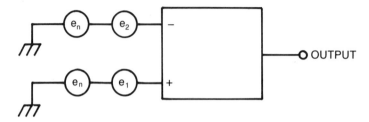

Fig. 10-41. Differential amplifier redrawn.

$$e_0 = A \left[e_1 + e_n - (e_2 + e_n) \right]$$

This can be rewritten as Equation 10-71.

Equation 10-71

$$e_0 = A \left[e_1 + e_n - e_2 - e_n \right]$$

As can be seen the output voltage equation then becomes:

$$e_0 = A \left[e_1 - e_2 \right]$$

The noise has therefore been canceled. Now, of course, noise developed in the amplifier itself will not be cancelled.

An example of a differential amplifier built using an op amp is shown in Fig. 10-42. This is simply an op amp wired for both equal gains for the noninverting and inverting input terminals.

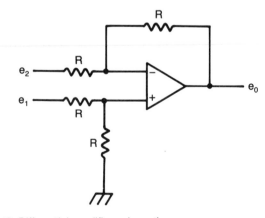

Fig. 10-42. Differential amplifier schematic.

SPECIAL FORMS OF NOISE

In Chapter 1 we discussed the various forms of noise. Noise, however, can be characterized in other ways. Some of these are as follows:

1. Noise dependent on the signal (signal related noise).
2. Idle network noise.
3. Impulse noise.
4. Crosstalk.

First of all, signal related noise occurs in a circuit whenever a signal is present. Such noise is prevalent in equipment such as equalization networks and decision circuits as appear in repeaters.

Idle Network Noise

This type of noise includes noise sources that produce noise in a rather consistent manner without a signal influence being present. In Chapters 1 and 3 we discussed the sources of such noise.

Also included in this noise is electromagnetic coupling. Almost all transmission networks are exposed to electrical disturbances that are caused by their proximity to other circuits. The ground loop is the epitome of such a coupling problem.

Echo is an undesirable condition that occurs on a telephone circuit. Depending on its strength, echo can range from nearly unnoticeable to a major annoyance. Echo is measured as the amount of loss in dB of reflected voice power. The larger the ERL (echo reflected loss) the less the amount of reflected energy coming back on the circuit, and the less echo that is heard. If ERL is reduced, the amount of reflected energy would be increased, causing poor quality.

There are several forms of echo. They may be listed as follows:

1. Talker Echo. In this situation the talker hears his/her own voice.
2. Singing Echo. Here, echo is amplified until a tone is produced due to feedback.

All circuits have this condition. It is especially noticeable if the return signal is high and circuit length is long. When the length is long, the time delay associated with reflected energy increases and serious echos occur.

For short circuit lengths, reflections return with small delay times and it will be noticed as a side tone. Echo effects can be cancelled by increasing the total end-to-end circuit loss so the total reflected energy can be decreased.

3. Listener Echo. Here the listener hears the talkers voice.

Common frequencies in the echo band are 500 to 2500 kHz.

Impulse Noise

This type of noise is characterized by large peaks, or impulses, in the overall waveform. Some of the sources of such noise are relay, motor, and switch operation, microwave path taking on carrier channels, and "cracking or popping" due to thermal effects in a transmission cable.

This type of noise has very little effect on the transmission of speech because the human ear can tolerate it. Data Communication Transmission is extremely sensitive to impulse noise. This is because data bits or groups of data can be completely destroyed, resulting in errors in the received signal. One of the most serious forms of impulse noise is that due to lightning.

Crosstalk

This type of noise is typical in pulse communication systems. It is the resultant of the frequency behavior of the channel or can be produced by inductive or capacitive coupling in parallel lines or at junctions of wires.

NOISE AND HUM MEASUREMENTS IN AMPLIFIERS

An oscilloscope can be used to make background noise checks on an amplifier. The vertical amplifier of the scope should be able to give a noticeable deflection with rather small input voltages such as one millivolt or smaller. There are, of course, many voltmeters that can do this but with the oscilloscope, we can see the nature of the noise and check for such things as hum and parasitic oscillations.

The procedure consists of the following steps. The amplifier in question is terminated in a load impedance equal to the amplifier output impedance. For example, a power amplifier that normally drives a 32-ohm speaker should be terminated in a resistance of 32 ohms. The oscilloscope is switched on internal sweep and internal sync. Any gain control is turned to maximum and tone controls to their normal point of operation. After this, the oscilloscope vertical gain control is adjusted until noise or has is indicated.

The first test that is made is to see just what is the source of the noise. The output from the amplifier is removed. If noise still stays on the scope, then we have noise being picked up from the outside world. If the noise has vanished, then we have noise from the amplifier.

Assuming that the noise did belong to the amplifier, you can measure the peak-to-peak amplitude of the total noise, which is composed of background noise, oscillation, hum and other disturbances. The amplifier is then switched to line sync. Should a stable signal pattern appear hum is present and its peak-to-peak amplitude is measured.

Should a frequency component be present that differs from the line frequency, oscillations are suspected. The amplifier input terminals are then shorted. Should this output signal disappear, the signal was probably due to pickup from external stray sources. Should the signal remain, it was most likely due to parasitic oscillations.

In Chapter 1, we noted that there are many sources of electrical noise. Lightning flashes both near and far are a source of rich broadband electromagnetic interference. Man-made devices such as ignition systems, electric motors, switches, relays, and diathermy machines are also sources of EM radiation. Power lines and transformers are also sources of interference which is periodic and low in frequency.

In electrical circuits, the "front end" of the circuit should be shielded by screens or metallic plates. The shield should be soldered in place for best results, however if hardware "screws and nuts" are used, the spacing of these screws should be very small compared to the frequency of the interference signal.

Any radiation interference at 60 Hz is frequently due to ground loops. A ground loop is defined as a closed loop of wire that should have all points at ground potential but does not, allowing induction of currents at normally low frequencies. These currents result in small but measurable voltages at the frequency of interference. To present this, bring all connectings to ground at the same point.

Electrical noise can also be created by mechanical vibrations. Vibrations range over a wide frequency range namely dc to several kHz. When a piece of electrical equipment sets in a room, a typical vibration may be on the order of fifteen hertz. It is only natural to believe that vibration frequency is inversely proportional to the object's size. The process in which mechanical vibration is converted to an electrical noise is microphonics. The noise that results is called *microphonic noise.*

Microphonic noise is very prevalent in vacuum tubes because vibration of the plate, cathode, and grids changes the interelectrode capacitance between the respective elements of the tube changing the output signal in a random manner. Solid-state devices are not susceptible to this due to the solid nature.

Whenever there is a cable, vibrations will result in noise signals being produced in the outer braided section. This is produced by variations in the contacts of the braid, dielectric changes, as well as effects similar to that which generate voltages in crystal materials. Here are some ways this may be prevented:

1. Eliminate the vibration.
2. Use shock absorbers.
3. Use a cable that is low in noise production.

INTERFERENCE REDUCTION

Protection against interference can be accomplished by proper grounding, careful circuit layout, and shielding. Shielding can be reduced to three principles:

1. The shield conductor should be connected to the zero reference of the signal (ground) only once.
2. The shield and signal reference ground should be grounded at the same point physically.
3. To be effective, all signal carrying conductors should be enclosed within the shield.

Circuit layout should avoid excessive crowding, long adjacent signal paths (with or without shielding) and unnecessary wire crossings.

ELECTROMAGNETIC INTERFERENCE (EMI)

This term relates to the condition where equipment and systems cannot work in their normal environment without interference between each other. There is also another condition labeled

320

(EMC), which is described as the situation that allows equipment and systems to operate together without interference and this is called electromagnetic compatibility.

SUMMARY

1. Noise output may be predicted for an amplifier by a calculation composed of various noise formulas.
2. Noise figure may be predicted by formulas which have been derived for different types of amplifier circuits.
3. Noise figure can be reduced in transistor voltages amplifiers by using either an optimum source resistance or an optimum collector current.
4. Noise is a serious problem also in integrated circuits. In digital integrated circuits noise margin and noise immunity are the important topics.
5. In a mobile system such as in an automobile, generator and ignition noise cause quite a big problem.
6. A fiber-optic link helps reduce noise during the transmission process.
7. The Dolby noise reduction system helps reduce hiss, hum and distortions which may re recorded on a cassette tape.

QUESTIONS AND PROBLEMS

1. Refer to Fig. 10-1. Suppose the following data is given:

$$R_g = 1 \text{ M}\Omega$$
$$R_s = 2 \text{ k}\Omega$$
$$R_{eq} = 4 \text{ k}\Omega$$
$$C = 100 \text{ pf}$$
$$R_L = 10 \text{ k}\Omega$$
$$\mu = 40.$$
$$R_d = 10 \text{ k}\Omega$$

Calculate noise output voltage and noise figure.

2. Refer to Fig. 10-5. Suppose the following data is given:

$$R_g = 1 \text{ M}\Omega$$
$$R_s = 2 \text{ k}\Omega$$
$$R_{eq} = 4 \text{ k}\Omega$$
$$R = 10 \text{ k}\Omega.$$

3. Refer to the common gate voltage amplifier shown in Fig. 10-6. Compute the noise figure if the following data is given:

$$R_s = 2 \text{ k}\Omega$$
$$R = 10 \text{ k}\Omega$$
$$R_{eq} = 4 \text{ k}\Omega.$$

4. For the circuit shown in Fig. 10-1 compute the optimum source resistance to minimize noise figure and calculate the turns ratio for the transformer required to do the matching.

5. Compute the minimum noise figure possible with varying collector current if the following data is given:

$$h_{FE} = 70$$
$$R_b = 200 \ \Omega$$
$$R_S = 100 \ \Omega.$$

6. What is the difference between noise gain and signal gain for an operational amplifier?

7. List the guidelines needed when designing an integrated circuit amplifier.

8. Compute the noise gain and the size of the compensating resistor for the circuit in Fig. 10-16B. $R2 = 100 \text{ k}\Omega$ and $R1 = 4 \text{ k}\Omega$.

9. List the causes for the following types of noise:

a. power supply spiking
b. induced noise
c. ground noise
d. capacitive coupled noise

10. Discuss some of the causes of noise in a radio for an automobile.

11. Discuss the causes of noise in an optical communication system and how a fiber-optic link can be used to lower its effect.

Chapter 11

Experiments

EXPERIMENT 1 NOISE AND HUM MEASUREMENTS ON AN AUDIO AMPLIFIER

Objective:

To perform an output noise and hum measurement on an audio amplifier.

Discussion:

Noise is generated in an amplifier by the passive and active circuit elements present. Hum is present in the output due to the line frequency and is generated by capacitive and inductive coupling of circuit elements to the internal power source or to external fields.

One way of measuring hum is to use an rms-reading ac voltmeter buffered from the amplifier output by a low pass filter. The filter is designed to pass the lower frequency harmonics of the line frequency which is strongest in amplitude. A certain amount of noise will be read by the meter since the meter must measure the noise within the filter bandwidth. An oscilloscope is usually the best tool for measurement of hum since you can measure the actual peak to peak waveform reading. Noise without hum may be measured with an rms-reading ac voltmeter isolated from the amplifier output by a high pass filter which removes the hum components.

Equipment:

Audio generator
Audio amplifier
Oscilloscope
True rms voltmeter or average reading-voltmeter.

Procedure:

1. Connect a resistor across the input terminals of the amplifier which is equal to the internal resistance of the transducer normally connected to the amplifier.
2. Connect a resistor across the output terminals of the amplifier which is equal to the amplifier output impedance.
3. Set the amplifier gain control at maximum and tone controls (if present) to their normal position.
4. Connect an oscilloscope across the output terminals and use a sensitivity setting which gives a reasonable noise indication. Measure the total noise voltage. Also measure the rms noise voltage with a voltmeter. If an average voltmeter is used, multiply the reading by 1.15 to obtain the true rms value.
5. Set the oscilloscope time base to the line position (if present). Should the oscilloscope pattern become still it indicates that sychronization is occurring with the line frequency and hum is present. Measure and record the hum amplitude.

Conclusion:

Write a personal conclusion to this experiment.

EXPERIMENT 2 RANDOM NOISE GENERATOR USING A ZENER DIODE

Objective:

To examine the operation of a noise generating device and observe the properties of random noise.

Discussion:

Random noise generators are available to cover frequencies from near dc to the microwave region. The method of generation

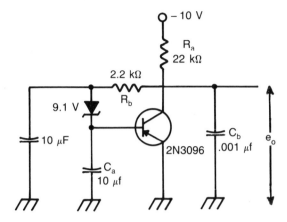

Fig. 11-1. Zener diode noise generator schematic.

is usually by amplification of the noise produced by a semiconductor noise diode.

In this experiment we will use a zener diode to produce a noise current, which will be amplified by a bipolar transistor. Typical noise currents of a zener diode are 20 to 40 microamperes. A frequency makeup of a zener diode may include frequencies up to about 300 kHz.

In Fig. 11-1 we see the zener diode noise generator. The zener is not only producing noise current but also a stabilization of the collector operating point.

Capacitor C_a serves as a bypass to ripple which may otherwise be amplified and appear in the output. Also, R_a and R_b serve as dc biasing components for the amplifier configuration. Capacitor C_b serves to filter the high end of the noise frequency spectrum.

Equipment:

> Low voltage regulated power supply
> Components to build generator
> Oscilloscope
> Audio VTVM

Procedure:

> 1. Build the noise generator as shown in Fig. 11-1, and apply power.

2. Vary the potentiometer until noise begins to appear on the oscilloscope.
3. Observe the output noise voltage on the oscilloscope and comment on the output waveshape. What is the highest noise peak in the output?
4. Measure the rms output voltage of the noise generator.
5. Place a .01 μF capacitor across the output terminals and record the effects on output voltage.

Questions

1. Explain the operation of the circuit in detail.
2. What changes would you make in the circuit to produce pink noise?
3. Why would a zener diode be used in a noise generator rather than a regular semiconductor diode?
4. Estimate the crest factor for the noise produced in this experiment. From this value would you say the noise is Gaussian in nature?

EXPERIMENT 3 NOISE GENERATOR INTEGRATED CIRCUIT

Objective:

To build a noise generator using an integrated circuit noise source.

Discussion:

National Semiconductor's MM 5837 digital pseudo-random sequence generator can be used to produce a broadband white noise signal which can be used in various audio applications. Extremely simple wiring for the integrated circuit makes it very attractive. By using a filter on the output of the circuit, it is possible to convert the white noise generator to a pink noise generator.

Equipment:

Dual Trace Oscilloscope
MM 5837 IC
Voltmeter

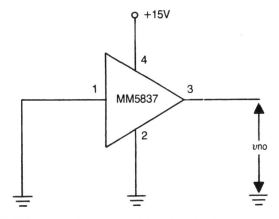

Fig. 11-2. Noise generator using an integrated circuit.

Procedure:

1. Connect the integrated circuit as shown in Fig. 11-2. Apply 15 volts dc to the circuit.
2. Examine the waveform at pin 3 of the integrated circuit with an oscilloscope. Attempt to make a rough sketch of the waveform. Record the highest noise amplitude present in the waveform.
3. Measure the rms voltage at the output of the circuit.
4. Connect the filter shown in Fig 11-3 across the output of the noise generator. What happens to the output waveform?

Questions:

1. Estimate the crest factor of the noise process.

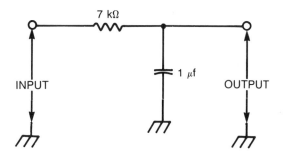

Fig. 11-3. Filter to place on noise generator.

2. Explain why the waveform looks smoother and makes less violent changes when the filter is connected as in procedure four.
3. If it were possible to remove all components from the noise waveform except one, what would the remaining waveform look like in its shape?

EXPERIMENT 4 TRANSISTOR
NOISE VOLTAGE GENERATOR

Objective:

To observe the operation of a noise generator (using a transistor as the noise source) by examination of the output waveform on an oscilloscope. Also, to filter the output waveform and listen to the noise on an audio amplifier.

Discussion:

Noise, with more or less "pure white" characteristics, can be filtered and shaped for various purposes, ranging from radio alignment, to music, or simulation of various natural sounds such as rain on the roof.

Many natural random impulse sources are available to the experimentor including the plasma from gas discharges occurring in various lamps as well as semiconductors biased into a state of noisiness. A zener diode is a very excellent noise source. However, under certain conditions a transistor with its base-to-emitter junction in the reverse biased mode can also be a very excellent noise source. If a large amount of amplification follows this noise source, the noise can be studied on an oscilloscope or a spectrum analyzer.

When observing noise that is very rich in frequency content, the time waveform takes on a very complex appearance that is difficult to record from an oscilloscope. By removing some of the frequencies using a filter, we can observe a more simple random wave on the scope.

Equipment:

Parts to construct noise generator (see Fig. 11-4)
Oscilloscope
Audio VTVM
Low Voltage Power Supply

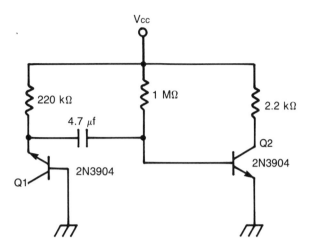

Fig. 11-4. Transistor noise voltage generator.

Procedure:

1. Construct the noise generator as shown in Fig. 11-4.
2. Increase the power supply voltage from zero to the value where noise begins to appear on the scope. Record the value where this occurs.
3. Continue to increase the supply voltage to 9 volts. Make a sketch of the scope presentation giving the approximate size of the highest voltage observed on the scope. Also measure the output noise rms using the audio VTVM.
4. Place a capacitor of 0.015 μF between collector of Q2 and ground. Measure the rms value of the noise using the audio VTVM.
5. Repeat the procedure four using capacitor values of 0.05 μF, 0.02 μF, and 0.01 μF.
6. Draw the output noise voltage waveform for the case when the capacitor is 0.01 μF.

Conclusions and Questions:

1. Explain in detail the operation of the transistor noise generator used in this experiment.
2. How many noise sources are present in the generator not counting the reverse biased transistor junction? What types of noise do the sources produce?

3. Why does an error occur from using an audio VTVM in this experiment?
4. As filtering capacitance is increased, what happens to the rms noise output voltage?
5. As filtering capacitance increases what changes occur in the noise waveform? Why?
6. How could you use a noise generator like the one in this experiment to help determine the frequency response of an amplifier?
7. Make a graph showing how noise output is varied by changing filtering capacitance. Plot voltage versus capacitance.
8. Estimate the noise equivalent bandwidth of this experiment for procedure 4.
9. Using the information of procedure 3, could you estimate if the noise is Gaussian?

EXPERIMENT 5 NOISE EQUIVALENT BANDWIDTH

Objective:

To learn how to determine noise equivalent bandwidth for an amplifier circuit.

Discussion:

The procedure for determining noise equivalent bandwidth graphically is given by the following two steps.

1. Determination of the transfer function characteristic and the area under the square of the transfer function.
2. Calculation of the noise equivalent bandwidth by a formula.

Equipment:

Parts to build amplifier shown
Two VTVMs
Audio Signal Generator
Batteries or power supplies

Procedure:

1. Construct the circuit as shown in Fig. 11-5.
2. Using an audio signal generator apply an input signal of

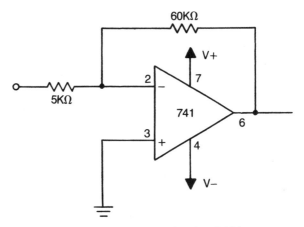

Fig. 11-5. Circuit to test for noise equivalent bandwidth.

.05 volt rms. Measure the output voltage for the frequencies shown in Fig. 11-6.

3. Determine the voltage gain for the circuit at each one of

FREQUENCY	VOLTAGE GAIN
20HZ	
50HZ	
100HZ	
150HZ	
200HZ	
300HZ	
500HZ	
700HZ	
1000HZ	
1200HZ	
1500HZ	
2000HZ	
3000HZ	
5000HZ	
7000HZ	
10,000HZ	
20,000HZ	
50,000HZ	
100,000HZ	
150,000HZ	
200,000HZ	

Fig. 11-6. Frequencies to use for Experiment 6.

the frequencies given in Fig. 11-6. Then square each voltage gain value.

4. Plot voltage gain squared versus frequency on linear graph paper.
5. Determine the area under the curve by breaking up the curve area into a group of rectangles and triangles. Call the area AR. Refer to Fig. 11-7 for an example of this procedure.
6. Determine the value of voltage gain squared where the frequency response is flat. Call this value $(A_p)^2$.
7. Calculate the noise equivalent bandwidth using this formula:

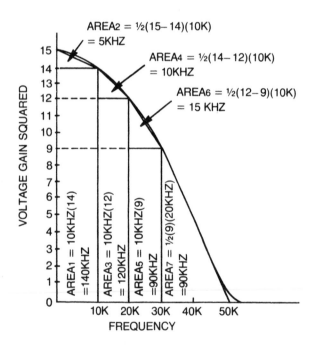

$$\text{NOISE EQUIVALENT BANDWIDTH} = \frac{\Sigma \text{AREAS}}{\text{MAX RESPONSE}}$$

$$B_{eq} = \frac{(140 + 5 + 120 + 10 + 90 + 15 + 90)\ \text{KHZ}}{15}$$

$$= \frac{470\ \text{KHZ}}{15} = 31.3\ \text{KHZ}$$

Fig. 11-7. Example of breaking area up into rectangles and triangles.

$$B_{eq} = \frac{\text{Area under Voltage Gain Squared Curve}}{(A_p)^2}$$

Conclusions:

1. Suppose the input spectral density to the amplifier in this experiment was white and equal to 8×10^{-8} volts²/hertz. What would the output noise voltage be?
2. In what condition would noise equivalent bandwidth be the same as the cutoff frequency of the network in question?
3. For the amplifier used in this experiment, what is the ratio of noise equivalent bandwidth to cutoff frequency?

EXPERIMENT 6 NOISE REDUCTION WITH FIBEROPTIC CABLE

Objective:

To observe noise reduction in a simple optical communication system by using a fiberoptic link.

Discussion:

Any communication system must contend with noise in the channel. For an optical system the channel may consist of the atmosphere or a fiberoptic cable. When using the atmosphere as the channel, we must deal with such things as precipitation, dust, and other light sources and scintillation. Losses are not easily predictable as with a fiberoptic link. Although a cable may have considerable loss, it does not have the random effects like that of the atmosphere. Perhaps the largest advantage is that the problem with external light sources does not exist.

In this experiment we will build an optical communciation system and observe the results of transmitting using the atmosphere and a fiberoptic link as the channel.

Equipment:

Transistor radio with earphone jack
Earphone cord
Lighting emitting diode
270 Ω resistor

Power amplifier
Two 1 1/2-volt batteries
Phototransistor
Fiberoptic cable

Procedure:

1. Connect the circuit as shown as in Fig. 11-8.
2. Turn the radio on and adjust the tuning dial and volume until the light-emitting diode (LED) begins to light.
3. Hold the phototransistor about two feet from the LED. Turn up the volume of the power amplifier until you hear the audio from the radio coming from the power amplifier. Alternately adjust the volume of the radio and amplifier until you obtain the best sound output.
4. Repeat steps two and three with Fig. 11-9. Record effects and notice the noise level in the output.
5. Connect the fiberoptic cable between the transmitter and receiver as shown in Fig. 11-10, and repeat procedure 4. Notice that the noise level is much smaller now. The fiberoptic cable is allowing only light from the LED to be transmitted. External light sources are prevented from influencing the output in this situation.

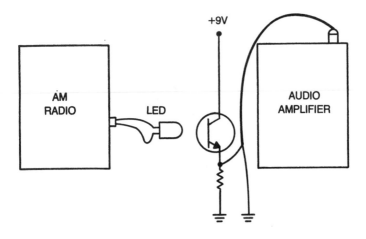

Fig. 11-8. Optical link without fiberoptic cable and LED connected directly to audio.

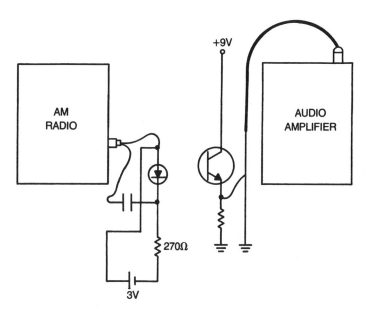

Fig. 11-9. Optical link without fiberoptic cable with LED modulator.

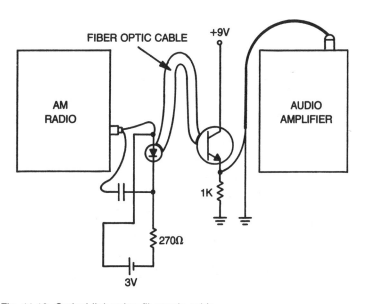

Fig. 11-10. Optical link using fiberoptic cable.

335

Conclusions:

1. Why does the fidelity seem improved when a dc bias is provided for the LED as in procedure four?
2. Explain the communication system operation completely. Why does the phototransistor require an external voltage source?
3. Explain why the noise level coming from the loudspeaker becomes smaller when using the fiberoptic link.

EXPERIMENT 7 ELECTRICAL NOISE IN A DC MOTOR

Objective:

To observe noise caused by an electric motor.

Discussion:

Whenever an electric spark occurs, many different frequency components are generated. These different frequency components may span a significant part of the frequency spectrum and interfere with the transmission of radio signals.

In this experiment we will observe the effect of noise on AM and FM transmissions.

Equipment:

Small dc motor 1 1/2 to 3 volts
Resistor decade box
3-volt battery
AM-FM radio

Procedure:

1. Connect the motor as shown in Fig. 11-11.
2. Hold the radio near the motor and tune the radio to an AM station. Observe the noise level in the speaker.
3. Repeat procedure two for an FM station.
4. Connect the decade box in series with the motor as in Fig. 11-12 and repeat steps two and three when the motor is going at a slow speed. Do this by adding resistance from the decade box which is just enough to allow the motor to still operate.

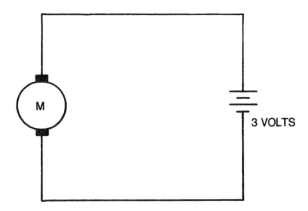

Fig. 11-11. Motor connected to battery directly.

Questions:

1. Why is the noise less noticeable from the FM condition than the AM condition?
2. What effects does decreasing motor speed have on the noise?
3. By what means does the motor noise become induced into the radio?

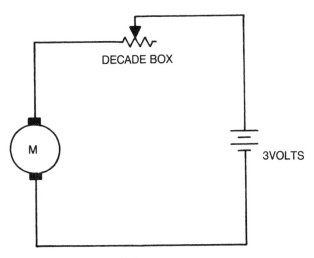

Fig. 11-12. Motor in series with battery.

EXPERIMENT 8 LOW-PASS
FILTER AND HIGH-PASS FILTER

Objective:

To construct and evaluate the operation of a low-pass and high-pass filter.

Discussion:

Regardless of how complicated a network might be, it will eventually produce a low-pass type response. This of course places a bandlimiting effect on any noise that is being produced. If the cut-off frequency is given by f_c and we have a single section low-pass filter, the noise bandwidth works out to be given by the following equation:

$$B_{eq} = 1.57 \, f_c$$

Because so many circuits have low-pass characteristics, it is only natural to evaluate such a circuit's frequency response. A frequency response is made up of two individual responses. One being the magnitude function, and the other, the phase function. In this experiment we will construct a basic low-pass and high-pass filter and perform a frequency response test on each one.

Equipment:

Parts to build filters (see Fig. 11-13)
Audio generator
VTVM's (2)

Procedure:

1. Construct the circuits shown in Fig. 11-13, and make a frequency response test on each. Gather information to plot the circuit responses for frequencies from 20 Hz to 200 kHz.
2. Measure the noise output voltage for each circuit with the generator set to zero output.
3. Plot the response of each network on semi-log paper.

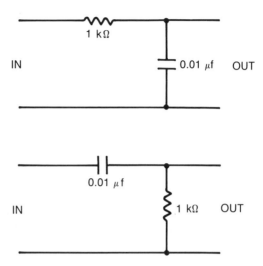

Fig. 11-13. Low-pass filter (top), high-pass filter (bottom).

Questions:

1. What is the cutoff frequency for each network?
2. Which network had the highest output noise voltage? Why?
3. Which of the following types of circuits would have the smallest output noise? Why?

   a. differentiator
   b. integrator

4. What is the noise equivalent bandwidth for a high-pass filter? Give the theoretical value.

EXPERIMENT 9 BANDPASS FILTER
Objective:

To construct a typical bandpass filter.

Discussion:

In Chapter 2 we discussed the basic types of filters that occur in electronic circuits. One type of filter characteristic that shows itself frequently in amplifiers is the bandpass type. Such a circuit places a bandpass type response on any white noise that may be present at the circuit's input terminals.

A bandpass filter may be constructed using both a low-pass and high-pass filter. If the two sections are isolated by a buffer or buffer-type amplifier, prediction of the overall response becomes easier than if no isolation exists. This is because the two sections load each other's impedance characteristic in the loaded condition.

In Fig. 11-14 we see the bandpass filter. Note that the input to the amplifier comes from the high-pass section. This is because the noninverting terminal of the op amp must be returned to ground through some dc path to set up a biasing current for the op amp's noninverting terminal.

Because the high-pass section has a resistor at this position, the condition is satisfied. The low-pass section is tied to the output terminal. Gain for the op amp is set by resistors R3 and R4. The gain for the amplifier is given by:

$$A_v = 1 + \frac{R3}{R4}$$

The high-pass section has its cutoff frequency equal to:

$$f_{c(low)} = \frac{1}{2\pi\, C_1\, R_1}$$

Note that we call the cutoff frequency $f_{c(low)}$. "Low" indicates that the frequency is the lowest cutoff frequency present. This is

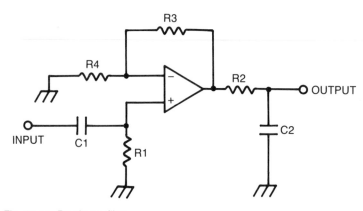

Fig. 11-14. Bandpass filter.

also the first cutoff frequency reached as frequency moves to the right of zero hertz. It is also the frequency where the gain of the circuit is 3 dB below the maximum gain.

The low-pass section has its cutoff frequency equal to:

$$f_{c(high)} = \frac{1}{2\pi\ C2\ R2}$$

Note that we call the cutoff frequency $f_{c(high)}$. "High" indicates that the frequency is the highest frequency present. This is the last cutoff frequency reached as frequency moves to the right from zero frequency. The high cutoff frequency is that frequency when the gain has dropped 3 dB from the maximum gain. Therefore combining the low- and high-pass sections produces a bandpass response.

Equipment:

Parts to construct filter (see Fig. 11-15)
Audio generator
Oscilloscope
Low-Voltage Power Supply
VTVM's (2)

Procedure:

1. Build the bandpass filter shown in Fig. 11-15.

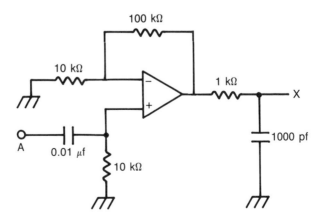

Fig. 11-15. Bandpass filter for procedure 1 in experiment 9.

2. Apply a sinewave of 0.1 volt rms between the input terminal and ground. Connect an oscilloscope between output and ground. Make sure that the output is a sinewave.
3. Keep the input voltage equal to 0.1 volt rms and vary the input frequency from 20 hertz to 200 kHz measuring the output voltage between point X to ground for the values in Fig. 11-15.
4. Calculate the gain for each of the tabulated frequencies. Use the following formula:

$$A_v = \frac{E_{out}}{E_{in}}$$

5. Remove the signal generator. Connect point A to ground. Measure the output noise voltage using an audio voltage meter.
6. Using semi-log graph paper, plot voltage-gain versus frequency for the amplifier used in this experiment.

Questions:

1. What was the 3 dB bandwidth for the amplifier used in this experiment? Do this two ways:

   a. by calculation
   b. by measurement

2. How close did the calculated bandwidth agree with the measured bandwidth?
3. Estimate the noise equivalent voltage of the amplifier using the maximum gain and the data of procedure 5.
4. Estimate the noise equivalent bandwidth for the amplifier using the relation:

$$B_{eq} = \frac{\pi}{2} \ (BW)$$

   Why is there an error in using this formula?

5. Estimate the noise equivalent resistance for the amplifier assuming a temperature of 20 °C.

6. Estimate the noise equivalent temperature for the amplifier if the source temperature were 20 °C and its internal resistance were 300 Ω.
7. Estimate the noise figure for the amplifier using the data of question 6.

EXPERIMENT 10 NOTCH FILTER

Objective:

To become familiar with a notch filter by construction and measurement of the frequency response.

Discussion:

Should a particular band of frequencies be causing some undesirable effect in a circuit, they can have their energy attenuated by a band-rejection filter. In this experiment we will construct a twin-tee network that will be built using resistors and capacitors. See Fig. 11-16 for an understanding of the circuit. The filter behaves as if it were a very small impedance connected to some output load when the frequency is very small or very large. At some intermediate frequency, the circuit appears as a very large impedance in series connected to some output load. Therefore, the transfer function for the band reject filter is near unity for low and high frequencies, but is near zero for the intermediate frequency.

This frequency where the transfer function is near zero is called the *null frequency*. The frequency at which this appears is given by:

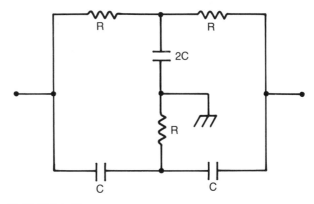

Fig. 11-16. Notch filter.

$$f_{null} = \frac{1}{2\pi\,R\,C}$$

Although the transfer function should be zero at the null, real world values normally work out to be an attenuation of about 30 dB for components with tolerances of about 20%.

Equipment:

Parts to construct filter (see Fig. 11-17)
Audio generator
Oscilloscope
VTVM's (2)

Procedure:

1. Construct the circuit shown in Fig. 11-17.
2. Run a frequency response test on the circuit gathering information to plot the voltage transfer function magnitude versus frequency.
3. Plot the response of the circuit versus frequency on semi-log graph paper.

Questions:

1. How close did the measured value of null frequency compare to the calculated value?

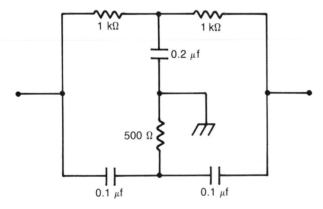

Fig. 11-17. Notch filter for procedure 1 in experiment 10.

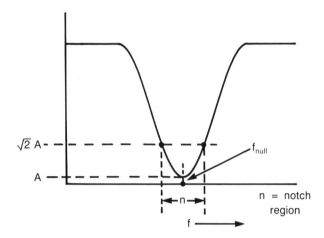

Fig. 11-18. Notch filter frequency characteristic.

2. How wide was the null region? Answer this by measuring the response of the circuit at the null. Multiply this value by the square root of 2 and find the two frequencies that define this value. Take the difference of those frequencies and the answer will be the null region. See Fig. 11-18 for an example.
3. How could you build a bandpass filter using an op-amp and a band reject filter?
4. List an application for a band reject filter in regards to noise.

EXPERIMENT 11
COMPRESSOR AND EXPANDER CIRCUITS

Objective:

To become familiar with compressor and expander circuits by plotting their characteristic.

Discussion:

A compressor is a circuit that offers a large voltage gain to small signals and a smaller voltage gain to larger signals. Such a characteristic can be useful in situations in which a distortion of the incoming signal is necessary such as in pulse-code modulation circuits. A circuit to produce compression can be constructed using an op

345

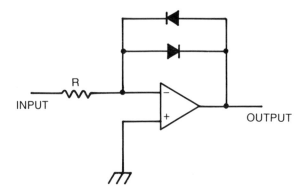

Fig. 11-19. Compressor.

amp, resistors, and diodes. The input signal amplitude determines the current through resistor R. By op-amp theory we know that this current must flow through the component bridged between the output terminal and the noninverting terminal of the op amp.

The voltage across a diode is related to the current flowing through it in a log type relationship. If the input voltage is large, the input current is large. The output voltage appearing across the diode is related to this current. If the input voltage is small, the input current is small. The output voltage is again related to this current. Because the logarithm of a large number is larger than the logarithm of a small number, there will be a larger output for the larger input signal than for the smaller input. However, the logarithm is not a linear operation and this rate of rise of output voltage drops as the input voltage increases. This then results in larger gains for small input signals than for large input signals.

Should the diodes be reversed in their circuit position we produce an expander circuit. Such a circuit produces small voltage gains for small inputs and large voltage gains for large input signals. In Fig. 11-19 we see a compressor schematic.

Equipment:

Parts to construct compressor
Audio generator
Oscilloscope
Low-Voltage Power Supply
VTVM's (2)

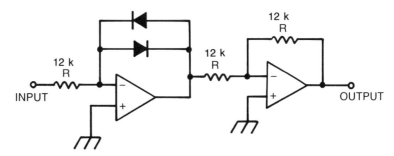

Fig. 11-20. Compressor for procedure 1 of experiment 11.

Procedure:

1. Build the circuit shown in Fig. 11-20. Use power supply voltages of ± 12 volts for each IC.
2. Gather information to plot v_{out} versus V_{in}. Do this by applying a small dc voltage to the V_{in} terminal. Gradually build up the dc voltage while you measure the dc output. You should measure V_{in} and V_o in both directions (plus and minus). Your plot should look something like Fig. 11-21.

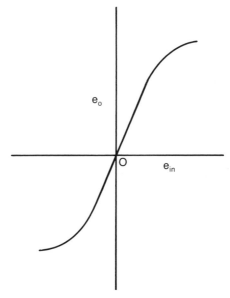

Fig. 11-21. Voltage characteristic for a compressor.

3. Apply a 1 kHz sinewave to the input. Using an input amplitude of 0.1-, 0.5-, and 1-volt peak, measure the circuit voltage gain.

Questions:

1. Describe the circuit operation as a compressor.
2. Show how the compressor circuit could be changed to an expander. Use a transistor instead of diodes.
3. Did the compressor used in this experiment have a logarithmic response?
4. Where are compressors used in the electronics field?
5. Show how you would construct a compressor and expander using a transistor.
6. What would be the effect of connecting a compressor in cascade with an expander?
7. What advantage is a compressor when dealing with noise?

EXPERIMENT 12 DESIGN OF TRANSISTOR VOLTAGE AMPLIFIER USING OPTIMUM COLLECTOR CURRENT

Objective:

To learn the procedure of design for minimum noise figure in a transistor voltage amplifier using optimum collector current.

Discussion:

Noise figure can be minimized in a transistor voltage amplifier by using either of the following techniques:

a. Source resistance optimization
b. Collector current optimization

In this experiment we will design by using optimum collector current. See Fig. 11-22 for the schematic.

Equipment:

One general-purpose transistor
Assortment of resistors
Audio generator
Single polarity low-voltage power supply
Oscilloscope

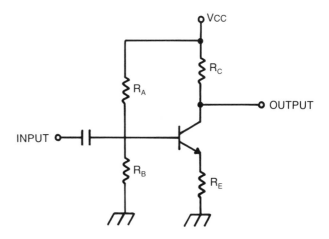

Fig. 11-22. Transistor voltage amplifier schematic.

Procedure:

1. Using the transistor, measure the h_{FE} on a curve tracer or transistor tester.
2. Measure the source resistance of the generator by the following method:

   a. Set the generator output voltage range to mid position. Measure the output voltage of the generator without a load. Call this voltage E_{oc} meaning "open circuit" output voltage.

   b. Adjust the decade box to maximum resistance and connect the decade box across the generator. Adjust the decade box until the output voltage of the generator is. $\dfrac{E_{oc}}{2}$ At this time the generator resistance is equal to the decade box resistance. Call the decade box resistance R_s.

3. Calculate the optimum collector current for the design.

$$I_{co} = \frac{\sqrt{h_{FE}}}{40\ R_s}$$

4. Pick a power supply voltage and call it Vcc.

5. Calculate the value of R_E that will allow one-tenth of the supply voltage to fall across R_E. Use the formula:

$$R_E = \frac{V_{CC}}{10\ I_{co}}$$

6. Make R_B 10 times the value of R_E. Make R_A 9 times R_B.
7. Design for a value of collector resistance that will allow one-half the difference of the supply voltage and V_E to appear across R_C.

$$R_C = \frac{0.45\ V_{CC}}{I_{co}}$$

8. The voltage gain of the circuit is then given by:

$$A_v = -\frac{R_C}{R_E}$$

assuming that R_E is unbypassed.

9. Run a frequency response test on the circuit. Gather information to plot gain-versus-frequency for frequencies as high as 100 kHz. Use the oscilloscope to make sure that the output is a sinewave.
10. Measure the output voltage with the generator voltage setting to zero.

Conclusions:

1. Assuming a base spreading resistance of 50 Ω, estimate the minimum noise figure (See Chapter 10).
2. What is the approximate input impedance for the circuit?
3. How would you rate the noise output of the circuit? (large, small). How is the noise output voltage related to the noise figure?

Appendix A

Solutions to Selected Problems

Chapter 1

Problem 5.

Thermal noise is related to temperature while shot noise is related to direct current flowing through an active device. Thermal noise occurs in resistive materials regardless if there is any current flowing from an outside source. Shot noise occurs only when a direct current is flowing and then only if it is the biasing current of an active device. Random emissions of electrons from an emitting surface give rise to the noise that has a noise current component, which is related to the square root of direct current, bandwidth, and a constant equal to twice the charge on the electron. Thermal noise is related to the square root of resistance, temperature, bandwidth, and four times Boltzmann's constant.

Both thermal and shot noise have a fairly flat spectrum in regards to noise-squared-voltage per frequency. Thermal noise closely resembles the true white noise spectrum and also has a Gaussian type probability function.

Problem 6.

A noise process is described as either white or pink in regards to the respective spectrum density plot. If a noise process has a flat type of spectral density it is then titled "white" since white light is said to be made of all frequencies or colors and this noise is made of all frequencies.

If the noise process has a spectral density that falls off as frequency rises then the density may be described as being pink. The word "pink" arose from the fact that red light is made of the lower frequencies or higher wavelengths and since noise that has a spectral density that falls with frequency, some components at all frequencies exist, so the noise is described as red and white together or pink.

Problem 7.

If bandwidth doubles a white noise will exhibit a doubling of power. For a pink noise process it is not as easy to explain. Suppose the bandwidth was defined by the frequencies 1 kHz and 2 kHz. This represents a 1 kHz bandwidth. Since pink noise is related to the natural log of the frequencies defining the bandwidth we can then write the following:

$$N_{pink1} \quad \alpha \quad l_n \frac{2 \text{ k}}{1 \text{ k}}$$
$$\alpha \quad 0.693$$

Suppose now we raise the bandwidth to 2 kHz. One way would be to increase the upper frequency limit to 3 kHz. The pink noise power would then be related to:

$$N_{pink2} \quad \alpha \quad l_n \frac{3 \text{ k}}{1 \text{ k}}$$
$$\alpha \quad 1.098$$

Therefore, raising the bandwidth by a factor of two would result in the noise power increasing by $\frac{1.098}{0.693} = 1.584$. Now suppose we repeat the problem for the case of a bandwidth of 1 kHz defined by the frequencies of 6 k and 7 kHz. The pink noise would now be given by:

$$N_{pink1} \quad \alpha \quad l_n \frac{7 \text{ k}}{6 \text{ k}}$$
$$\alpha \quad 0.154$$

If we increase the bandwidth by a factor of two by raising the upper frequency limit to 8 kHz results in a noise power of:

$$N_{pink2} \quad \alpha \quad l_n \frac{8 \text{ k}}{6 \text{ k}}$$
$$\alpha \quad 0.287$$

Therefore, raising the bandwidth by a factor of two in this case results in an increase in noise power of $\frac{0.287}{0.154} = 1.87$. We therefore conclude that we cannot define exactly what happens to the pink noise power when bandwidth raises by a factor of two except to say that there is some increase in noise power.

Problem 8.

$$E_n^{\,2} = (\text{Spectral Density})\ (\text{Bandwidth})$$

$$E_n^{\,2} = \left(7 \times 10^{-8} \frac{V^2}{Hz} \right) (1 \times 10^5 \ Hz)$$

$$E_n^{\,2} = 7 \times 10^{-3} \ V^2$$

$$E_n^{\,2} = 70 \times 10^{-4} \ V^2$$

$$E_n = 8.37 \times 10^{-2} \ \text{volts Or } 83.7 \text{ millivolts}$$

Problem 9.

$$E_n^{\,2} = K \, 1_n \frac{f2}{f1}$$

$$E_n^{\,2} = 8 \times 10^{-9} \ l_n \frac{2400}{600}$$

$$E_n^{\,2} = 8 \times 10^{-9} \ l_n \, 4$$

$$E_n^{\,2} = 8 \times 10^{-9} \ (1.386)$$

$$E_n^{\,2} = 11.09 \times 10^{-9} = 110.9 \times 10^{-10}$$

$$E_n = 10.53 \times 10^{-5} \ \text{volts or } 105.3 \ \mu V$$

Problem 10.

$$\text{Mean Value } \overline{V} = \frac{V_1\,n_1 + V_2\,n_2 + V_3\,n_3 + V_4\,n_4 + V_5\,n_5}{n}$$

In this problem the mean value is determined by taking each set of a piece of data, adding it with the other data values and dividing by the number of pieces of data. In this problem each value occurs only once so we have the following:

$$n_1 = n_2 = n_3 = n_4 = n_5$$

$$\overline{V} = \frac{100.7 + 111.9 + 105.6 + 99.4 + 95.6}{5}$$

$$\overline{V} = \frac{513.2}{5} = 102.4$$

Variance

$$V = \frac{(V_1 - \overline{V})^2 + (V_2 - \overline{V})^2 + \ldots}{n}$$

We first must find each deviation:

$$V_1 - \overline{V} = 100.7 - 102.64 = -1.94$$
$$V_2 - \overline{V} = 111.9 - 102.64 = 9.26$$
$$V_3 - \overline{V} = 105.6 - 102.64 = 2.96$$
$$V_4 - \overline{V} = 99.4 - 102.64 = -3.24$$
$$V_5 - \overline{V} = 95.6 - 102.64 = -7.04$$

We then apply the formula:

$$V = \frac{(-1.94)^2 + (9.26)^2 + (2.96)^2 + (-3.24)^2 + (-7.04)^2}{5}$$

$$= \frac{3.76 + 85.75 + 8.76 + 10.5 + 49.6}{5}$$

$$= \frac{158.37}{5} = 31.67$$

We then determine the standard deviation by taking the square root of the variance:

$$\sigma = \sqrt{V} = \sqrt{31.67} = 5.63$$

Problem 11.

The Gaussian curve or as sometimes called the "Bell curve" shows the probability of various instantaneous noise voltages occurring. The vertical axis shows spot probability while the horizontal axis shows the instantaneous noise voltage. Sometimes the center of the horizontal axis is zero, while other times it may be some value other than zero. The deviations about the mean are more likely to be small than large. If you examine the probability curve closely, you can see that the curve falls off either side of the mean. The farther you move from the mean, the smaller the spot probability or the less likely the noise voltage. We also note that the probability curve is a function of the horizontal axis variable squared.

Problem 12.

For the rising portion of the waveform, we may write the equation for a ramp.

Since the waveform rises to a peak in 1-2 μsec, we may write

$$y = mx + b$$

where

y is the voltage

m is the slope

x is the time value

b is the y intercept which is zero in this example.

We may then write,

$$y = \frac{\text{rise}}{\text{run}} t$$

This means that at 1.2×10^{-6} sec the waveform is V. The constant a must be evaluated.

From the model we know that the waveform must reach $\frac{V}{2}$ at 50 μsec.

Therefore we write

$$\frac{V}{2} = V \epsilon^{-\left(50 \times 10^6 - 1.2 \times 10^{-6}\right)a}$$

$$0.5 = \epsilon^{-48.8 \times 10^{-6}a}$$

or,

$$2 = \epsilon^{48.8 \times 10^{-6}a}$$

By definition, $48.8 \times 10^{-6}a$ is the natural log of 2.

$$y = \frac{V}{1.2 \times 10^{-6}} t$$

$$V(t) = 0.833 \times 10^6 \text{ V } t$$

$$V(t) = 8.33 \times 10^5 \text{ V } t$$

This equation is valid between zero and 1.2×10^{-6} sec.

Now, at 1.2 μsec the voltage begins to fall and reaches $\frac{V}{2}$ at 50 μsec. It then decays to zero at infinity.

We may write, the equation for an exponential decay,

$$V)t) = V \epsilon^{-(t - 1.2 \times 10^{-6})a}$$

evaluation at 50 μsec we have,

$$\ln 2 = 0.69$$

$$.69 = 48.8 \times 10^{-6}a$$

$$a = \frac{6.9 \times 10^{-1}}{4.88 \times 10^5}$$

$$= 1.4 \times 10^4$$

Therefore, the equation for the waveform from 1.2 μsec to ∞ becomes:

$$V(t) = V\Sigma^{-(t - 1.2 \times 10^{-6})\, 1.4 \times 10^4}$$

Problem 13.

Solar activity is occurring to some degree at all times. When storms on the sun increase, large amounts of charged particles are hurled towards outer space. Those particles that come into the vicinity of the earth are forced into the region of the north and south poles where they cause various gases in the earth's ionosphere to become ionized giving off various colors of light. This activity interferes seriously with radio reception.

Problem 14.

Cosmic noise is a form of space noise that is caused by stars other than our sun. Solar noise is a form of space noise that is caused by our sun. Solar noise is of interest from about 7 MHz to more than 1 GHz.

Problem 15.

Energy* Content per stroke is about 400 horsepower. Therefore the power dissipated may be:

$$(400 \text{ horsepower}) \left(746 \ \frac{\text{Watts}}{\text{horsepower}}\right)$$

$$= 298.4 \text{ Kilowatts.}$$

*The word energy is used since it is a popular term. Horsepower is not a unit of energy but of energy per unit time.

Problem 16.

$$V = \frac{K\,Q}{D}$$

$$= \frac{(9 \times 10^9)\,(16 \times 10^{-6})}{12 \times 10^{-2}}$$

$$= 12 \times 10^5$$

$$= 1.2 \times 10^6 \text{ volts}$$

Chapter 2

Problem 2.

The waveform can be defined as follows:

$$V_{(t)} = 40 \text{ volts} \qquad 0 \leq t \leq 1 \text{ sec.}$$
$$= 0 \qquad\qquad 1 \text{ sec.} \leq t \leq 2 \text{ sec.}$$

$$E_{av} = \frac{\displaystyle\int_0^T e_{(t)} dt}{T}$$

$$= \frac{\displaystyle\int_0^{1 \text{ sec}} 40 \, dt}{2 \text{ sec}} = \frac{40 \, t \, \Big|_0^1}{2}$$

$$= 20 \, t \, \Big|_0^1 = 20 \text{ volts}$$

$$E_{rms} = \sqrt{\frac{\displaystyle\int_0^T (V_{(t)})^2 dt}{T}}$$

$$= \sqrt{\frac{\displaystyle\int_0^1 (40)^2 dt}{2}}$$

$$= \sqrt{\frac{1600 \, t \, \Big|_0^1}{2}}$$

$$= \sqrt{800\,t\ \Big|_0^1}$$

$$= \sqrt{800} \quad = 28.28 \text{ volts}$$

Problem 3.

$$\text{Form Factor} = \frac{\text{rms}}{\text{Average}}$$

$$= \frac{28.28 \text{ volts}}{20 \text{ volts}} = 1.414$$

$$\text{Crest Factor} = \frac{\text{Peak}}{\text{rms}} \,.$$

$$= \frac{40 \text{ volts}}{28.28 \text{ volts}} = 1.414$$

Problem 4.

First we must define the waveform:

$$v_{(t)} = \frac{\text{Rise}}{\text{Run}}\ t = \frac{.4}{.1}\ t = 4\,t \quad 0 \le t \le .1 \text{ sec}$$

The dc component can be evaluated as follows:

$$A_0 = \frac{1}{T} \int_0^{\frac{T}{2}} f_{(t)}\ dt$$

$$A_0 = \frac{1}{0.1} \int_0^{.1} 4t\ dt = 10 \left[\frac{4t^2}{2} \right]_0^{.1}$$

$$= 5\,[4t^2]^{10} = 20t^2 = 20\,[0.1]^2 = .2 \text{ volts}$$

Problem 5.

To evaluate the transfer function we can use the voltage divider theorem for the first two circuits.
For Network A:

$$\frac{E_o}{E_i} = \frac{\dfrac{1}{jwc}}{jwL + \dfrac{1}{jwc}}$$

If we multiply numerator and denominator by jwc we obtain:

$$\frac{E_o}{E_i} = \frac{1}{-w^2LC + 1}$$

For Network B:

$$\frac{E_o}{E_i} = \frac{R}{R + jwL + \dfrac{1}{jwc}}$$

If we multiply numerator and denominator by jwc we obtain:

$$\frac{E_o}{E_i} = \frac{jwcR}{jwcR - w^2LC + 1}$$

For Network C:

1. Write the loop equations:

$$E_i = I_1 \, (jwL) + I_1 \, (R_1) - I_2 \, (R_1)$$

$$E_o = -I_1 \, (R_1) + I_2 \, (R_1) + I_2 \left(\frac{1}{jwc}\right) + I_2 \, (R_2)$$

2. Find Δ

$$\Delta = \begin{vmatrix} jwL + R_1 \, , & -R_1 \\[2mm] -R_1 \, , & R_1 + R_2 + \dfrac{1}{jwc} \end{vmatrix}$$

$$= (jwL + R_1) \, (R_1 + R_2 + \frac{1}{jwc}) - (-R_1) \, (-R_1)$$

$$= jwL \, R_1 + jwL \, R_2 + \frac{L}{C} + R_1 R_2 + \frac{R_1}{jwc}$$

3. Find ΔI_2

$$\Delta I_2 = \begin{vmatrix} jwL + R_1 & , & E_i \\ -R_1 & , & 0 \end{vmatrix}$$

$$= (jwL + R_1) \, (0) - (-R_1) \, E_i$$

$$= R_1 \, E_i$$

Find I_2

$$I_2 = \frac{\Delta I_2}{\Delta} = \frac{R_1 \, E_i}{jwL \, R_1 + jwL \, R_2 + \dfrac{L}{C} + R_1 \, R_2 + \dfrac{R_1}{jwc}}$$

Find E_o

$$E_o = I_2 \, R_2 = \frac{R_1 R_2 \, E_i}{\Delta}$$

Find $\dfrac{E_o}{E_i}$

$$\frac{E_o}{E_i} = \frac{R_1 \, R_2}{jwL \, R_1 + jwL \, R_2 + \dfrac{L}{C} + R_1 \, R_2 + \dfrac{R_1}{jwc}}$$

If we multiply numerator and denominator by jwc:

$$\frac{E_o}{E_i} = \frac{jwc \, R_1 \, R_2}{-w^2 CL \, R_1 - w^2 \, CL \, R_2 + jwL + jwc \, R_1 \, R_2 + R_1}$$

Problem 6.

The magnitude of the transfer function for Equation 2-18 is determined as follows:

$$\frac{E_o}{E_i} = \frac{1}{1 - w^2 c^2 R^2 + j3wcR}$$

$$\left| \frac{E_o}{E_i} \right| = \left| \frac{1}{1 - w^2 c^2 R^2 + j3wcR} \right|$$

$$\left| \frac{E_o}{E_i} \right| = \frac{1}{\sqrt{(1 - w^2 c^2 R^2)^2 + (3wcR)^2}}$$

$$\left| \frac{E_o}{E_i} \right| = \frac{1}{\sqrt{1 - 2 \, w^2 c^2 R^2 + w^4 c^4 R^4 + 9 \, w^2 c^2 R^2}}$$

$$\left| \frac{E_o}{E_i} \right| = \frac{1}{\sqrt{1 + 7 \, w^2 c^2 R^2 + w^4 c^4 R^4}}$$

The magnitude of the transfer function for Equation 2-19 is determined as follows:

$$\frac{E_o}{E_i} = \frac{A}{1 + jwcR}$$

$$\left| \frac{E_o}{E_i} \right| = \left| \frac{A}{1 + jwcR} \right|$$

$$\left| \frac{E_o}{E_i} \right| = \frac{A}{\sqrt{1 + (wcR)^2}}$$

The magnitude of the transfer function for Equation 2-20 is determined as follows:

$$\frac{E_o}{E_i} = \frac{1}{1 + j2wcR - w^2 c^2 R^2}$$

$$\left| \frac{E_o}{E_i} \right| = \left| \frac{1}{1 + j2wcR - w^2 c^2 R^2} \right|$$

$$\left| \frac{E_o}{E_i} \right| = \frac{1}{\sqrt{(1 - w^2c^2R^2)^2 + (2\ wcR)^2}}$$

$$\left| \frac{E_o}{E_i} \right| = \frac{1}{\sqrt{1 - 2w^2c^2R^2 + w^4c^4R^4 + 4w^2c^2R^2}}$$

$$\left| \frac{E_o}{E_i} \right| = \frac{1}{\sqrt{1 + 2w^2c^2R^2 + w^4c^4R^4}}$$

Problem 7.

The duty cycle should be given as follows: The pulse lasts for 6 msec. and the period is 10 msec. The duty cycle is then $\frac{6}{10} = 0.6$

Problem 8.

The average level is:

$$\begin{aligned} \text{Average level} \quad &= \text{V (duty cycle)} \\ &= 20\ (0.6) \\ &= 12\ \text{volts} \end{aligned}$$

The rms level is given by

$$\text{rms level} = \sqrt{\frac{\text{Area under Waveform squared}}{\text{period}}}$$

$$\begin{aligned} \text{rms} &= \sqrt{\frac{(20)^2\ (6)}{10}} \\ &= \sqrt{\frac{400\ (6)}{10}} \\ &= \sqrt{240} \\ &= 15.5\ \text{volts} \end{aligned}$$

363

Problem 9.

a) Square Wave period:

$$T = \frac{1}{f} = \frac{1}{\frac{\omega}{2\pi}}$$

$$= \frac{1}{\frac{1000}{6.28}} = 6.28 \text{ msec.}$$

b) See Fig. 2-49.
c) See Fig. 2-50.

Problem 10.
   See Fig. 2-51.

Problem 14.

$$\text{Substitute 5 radians/sec for } w$$

$$|Hc;jw,| = \frac{10}{\sqrt{(5)^4 + 20(5)^2 + 64}}$$

$$= \frac{10}{\sqrt{625 + 500 + 64}}$$

$$= \frac{10}{\sqrt{1189}}$$

$$= \frac{10}{34.4} = 0.29$$

$$= -\arctan \frac{6(5)}{8 - (5)^2}$$

$$= -\arctan \frac{30}{8\text{-}25}$$

$$= - \arctan \frac{30}{-17}$$

$$= - 180 - \arctan \frac{30}{17}$$

$$= - 180 - 60.5$$

$$= - 119.5°$$

Determine the magnitude and phase at $w = 5$ rad/sec for a transfer function given by

$$Hc;w = \frac{10}{-w^2 + 6;w + 8}$$

The magnitude is:

$$|Hc;w| = \frac{10}{\sqrt{(8 - w^2)^2 + (6w)^2}}$$

$$= \frac{10}{\sqrt{64 - 16w^2 + w^4 + 36w^2}}$$

$$= \frac{10}{\sqrt{w^4 + 20w^2 + 64}}$$

The phase is:

$$= \arctan \frac{0}{10} - \arctan \frac{6w}{8 - w^2}$$

$$= 0 - \arctan \frac{6w}{8 - w^2}$$

Problem 15.

$$v(\tau) = V \sin w\tau \quad 0 \leq \tau \leq \frac{T}{2}$$

$$v(\tau) = -V \sin w\tau \quad \frac{T}{2} \leq \tau \leq T$$

1. Find the average value of the waveform:

$$A_o = \frac{1}{T} \int_0^T v(\tau)^{dt}$$

$$= \frac{1}{T} \int_0^{T/2} V \sin wt \, dt + \frac{1}{T} \int_{T/2}^T -V \sin wt \, dt$$

$$= \frac{V}{T} \int_0^{T/2} \sin wt \, dt - \int_{T/2}^T \frac{V}{T} \sin wt \, dt$$

We note that $w = \frac{2\pi}{T}$

where T = period.

$$A_o = \frac{V}{T} \left[\frac{-\cos \frac{2\pi}{T} t}{\frac{2\pi}{T}} \right]_0^{\frac{T}{2}} - \frac{V}{T} \left[\frac{-\cos \frac{2\pi}{T} t}{\frac{2\pi}{T}} \right]_{\frac{T}{2}}^T$$

$$= \frac{V}{2\pi} \left[-\cos \frac{2\pi}{T} \left(\frac{T}{2} \right) + \cos \frac{2\pi}{T} (0) \right]$$

$$- \frac{V}{2} \left[-\cos \frac{2\pi}{T} \; (T) + \cos \frac{2\pi}{T} \left(\frac{T}{2} \right) \right]$$

$$= \frac{V}{2} \left[-\cos \pi + \cos 0 \right] \frac{-V}{2\pi} \left[-\cos 2\pi + \cos \pi \right]$$

$$= \frac{V}{2} \left[-(-1) + 1 \right] \frac{-V}{2\pi} \left[-1 - 1 \right]$$

$$= \frac{V}{2\pi} [2] + \frac{V}{2\pi} [2]$$

$$A_o = \frac{2V}{\pi}$$

2. Find the An coefficients:

$$A_n = \frac{2}{T} \int_0^T v(t) \; \cos \left(\frac{2\pi n}{T} \right) dT$$

$$= \frac{2}{T} \int_0^{\frac{T}{2}} \left(V \sin \frac{2\pi}{T} T \right) \left(\cos \frac{2\pi n}{T} T \right) dT$$

$$+ \frac{2}{T} \int_{\frac{T}{2}}^T \left(-V \sin \frac{2\pi}{T} T \right) \left(\cos \frac{2\pi n}{T} T \right) dT$$

We know the identity:

$$(\sin A)(\cos B) = \frac{\sin \; (A+B) \; + \; \sin \; (A-B)}{2}$$

Therefore:

$$A_n = \frac{V}{T} \int_0^{\frac{T}{2}} \left[\sin \frac{2\pi}{T}(1+n)T + \sin \frac{2\pi}{T}(1-n)T \right] dT$$

$$- \frac{V}{T} \int_{\frac{T}{2}}^{T} \left[\sin \frac{2\pi}{T}(1+n)T + \sin \frac{2\pi}{T}(1-n)T \right] dT$$

$$A_n = \frac{V}{T} \left[\frac{-\cos \frac{2\pi}{T}(1+n)T}{\frac{2\pi}{T}(1+n)} \right]_0^{\frac{T}{2}} + \frac{V}{T} \left[\frac{-\cos \frac{2\pi}{T}(1-n)T}{\frac{2\pi}{T}(1-n)} \right]_0^{\frac{T}{2}}$$

$$- \frac{V}{T} \left[\frac{-\cos \frac{2\pi}{T}(1+n)T}{\frac{2\pi}{T}(1+n)} \right]_{\frac{T}{2}}^{T} - \frac{V}{T} \left[\frac{-\cos \frac{2\pi}{T}(1-n)T}{\frac{2\pi}{T}(1-n)} \right]_{\frac{T}{2}}^{T}$$

$$A_n = \frac{V}{2\pi(1+n)} \left[-\cos \pi(1+n)+1 \right] + \frac{V}{2\pi(1-n)} \left[-\cos \pi(1-n)+1 \right]$$

$$- \frac{V}{2\pi(1+n)} \left[-\cos 2\pi(1+n) + \cos \pi(1+n) \right]$$

$$- \frac{V}{2\pi(1-n)} \left[-\cos 2\pi(1-n) + \cos \pi(1-n) \right]$$

For odd values of n all the A_n coefficients produced by the above equation give zero. For even values of n the A_n coefficients become equal to:

$$A_n = \frac{2\ V}{\pi\ (1+n)} + \frac{2\ V}{\pi\ (1-n)}$$

$$= \frac{2\ V}{\pi} \left[\frac{1}{1\ +\ n} + \frac{1}{1\ -\ n} \right]$$

$$= \frac{2\ V}{\pi} \left[\frac{1-n\ +\ 1+n}{(1+n)\ (1-n)} \right]$$

$$= \frac{2\ V}{\pi} \left[\frac{2}{1-n^2} \right]$$

$$A_n = \frac{4\ V}{\pi\ (1\ -\ n^2)}$$

We can change this expression to

$$A_n = \frac{4\ V}{-\ \pi\ (n^2-1)}$$

Therefore, even harmonic cosine waves do exist.

3. Find the B_n coefficients

$$b_n = \frac{2}{T} \int_O^T V(t) \sin\left(\frac{2\pi\ n}{T} T\right) dt$$

$$= \frac{2}{T} \int_O^{\frac{T}{2}} V \left(\sin\frac{2\pi}{T} T \right)\left(\sin\frac{2\pi\ n}{T} T \right) dT$$

$$+ \frac{2}{T} \int_{\frac{T}{2}}^{T} -V \left(\sin\frac{2\pi}{T} T \right)\left(\sin\frac{2\pi\ n}{T} T \right) dT$$

This leads to a final value of zero for all values of the integer n. Therefore, no sinewave components exist.

    4. Write the Fourier series as follows: We know all B_n values are zero.

$$v_{(t)} = A_o + \sum_{n=1}^{\infty} A_n \cos \frac{2\pi n}{T} t$$

The average level is $A_o = \dfrac{2\ V}{\pi}$

The 2nd harmonic occurs when n = 2 the signal would be

$$- \frac{4\ V \cos}{\pi\ (n^2 - 1)} \frac{2\pi n}{T} \ T \ \text{where n = 2.}$$

or,

$$- \frac{4\ V}{\pi\ (3)} \cos \frac{2\pi(2)T}{T} = \frac{4\ V}{3\pi} \cos \frac{4\pi T}{T}$$

In a like manner,

$$A4 = \frac{-\ 4\ V}{\pi\ (15)} \cos \frac{8\pi}{T} t$$

Therefore, the Fourier series would be:

$$v(t) = \frac{2\ V}{\pi} - \frac{4\ V}{3\pi} \cos \frac{4\pi}{T} t - \frac{4V}{15\pi} \cos \frac{8\pi}{T} t$$

$$- \frac{4\ V}{35\pi} \cos \frac{12\pi}{T} t - \frac{4\ V}{63\pi} \cos \frac{16\pi}{T} t$$

Problem 4.

    To work this problem it may make the evaluation simpler if sine and cosine terms are exchanged for Euler's equations. For example:

$$\cos x = \frac{e^{jz} + e^{-jz}}{2}$$

and

$$\sin x = \frac{e^{jz} - e^{-jz}}{2j}$$

In these problems, z is equal to $\dfrac{2\pi n}{T}$ we have cos

$$\cos \frac{2\pi n}{T} \quad \frac{e^{\,j\,\frac{2\pi n}{T}} + e^{-j\,\frac{2\pi n}{T}}}{2}$$

$$\sin \frac{2\pi n}{T} = \frac{e^{\,j\,\frac{2\pi n}{T}} - e^{-j\,\frac{2\pi n}{T}}}{2j}$$

1. We now proceed with defining the function.

$$v(t) = \frac{Rise}{Run} \quad T = \frac{0.4}{0.1} \quad T = 4T$$

$$0 \leq T \leq 0.1 \ sec$$

The dc component can be evaluated as follows:

$$A_0 = \frac{1}{T} \int_0^T f_{(t)} \, dT$$

$$= \frac{1}{0.1} \int_0^{0.1} 4T \, dT = 10 \left[\frac{4T^2}{2} \right]_0^{0.1}$$

$$= 5 \, [4T^2] \, \Big|_0^{0.1} = 20T^2$$

$$= 20 \quad 0.1^2 = 0.2 \ volts$$

Now, we find the A_n coefficients

2. $\quad A_n = \dfrac{2}{T} \displaystyle\int_0^T (mT) \, \cos \dfrac{2\pi nT}{T} \, dT$

$\quad = \dfrac{2m}{T} \displaystyle\int_0^T T \, \cos \dfrac{2\pi nT}{T} \, dT$

$\quad = \dfrac{2\,m}{T} \displaystyle\int_0^T T \left[\dfrac{e^{j\,\frac{2\pi nT}{T}} + e^{-j\,\frac{2\pi nT}{T}}}{2} \right] dT$

$\quad = \dfrac{m}{T} \displaystyle\int_0^T T \, e^{j\frac{2\pi nT}{T}} \, dT$

$\quad + \dfrac{m}{T} \displaystyle\int_0^T T \, e^{-j\frac{2\pi nT}{T}} \, dT$

Now, we can apply the exponential integral:

$$\int x \, e^{ax} \, dx = \dfrac{e^{ax}}{a^2} \, (ax - 1)$$

$$\text{Let } T = x, \quad a = j\,\dfrac{2\pi nT}{T}$$

Then,

$$\dfrac{m}{T} \int_0^T t \, e^{j\frac{2\pi nt}{T}} \, dt$$

$$= \dfrac{m}{T} \left[\dfrac{e^{j\frac{2\pi nt}{T}}}{\left(j\frac{2\pi n}{T} \right)^2} \left(j\frac{2\pi n}{T} t - 1 \right) \right]_0^T$$

$$= \frac{m}{T}\left[\frac{e^{j2\pi n}}{-\left(\dfrac{2\pi n}{T}\right)^2}(j2\pi n - 1) - \frac{1}{\left(\dfrac{2\pi n}{T}\right)^2}\;(\;1)\right]$$

Let T = x, $\quad a = -j\;\dfrac{2\pi n}{T}$

Then,

$$\frac{m}{T}\int_{0}^{T} t\,e^{-j\frac{2\pi nt}{T}}\,dT$$

$$= \frac{m}{T}\left[\frac{e^{-j\frac{2\pi nt}{T}}}{\left(-j\,\dfrac{2\pi n}{T}\right)^2}\left(-j\,\frac{2\pi n}{T}\;t - 1\right)\right]_{0}^{T}$$

$$= \frac{m}{T}\left[\frac{e^{-j\frac{2\pi n}{T}}}{-\left(\dfrac{2\pi n}{T}\right)^2}\;(-j\,2\pi n - 1)\right.$$

$$\left. - \frac{1}{-\left(\dfrac{2\pi n}{T}\right)^2}\;(\;1)\right]$$

$$A_n = \frac{m}{T}\left[\frac{(-)\,e^{j2\pi n}}{-\left(\dfrac{2\pi n}{T}\right)^2}\;(j2\pi n)\right.$$

$$+ \frac{e^{j2\pi n}}{-\left(\dfrac{2\pi n}{T}\right)^2} - \frac{1}{\left(\dfrac{2\pi n}{T}\right)^2}$$

$$+ \frac{e^{-j2\pi n}}{\left(\dfrac{2\pi n}{T}\right)^2}\;(j2\pi n)$$

$$+ \frac{e^{-j\pi n}}{\left(\dfrac{2\pi n}{T}\right)^2} - \frac{1}{\left(\dfrac{2\pi n}{T}\right)^2}\Bigg]$$

$$A_n = \frac{m}{T} \left[\frac{-e^{j2\pi n} + e^{-j2\pi n}}{\left(\dfrac{2\pi n}{T}\right)^2} \right]$$

$$+ \frac{m}{T} \left[\frac{e^{j2\pi n} + e^{-j2\pi n}}{\left(\dfrac{2\pi n}{T}\right)^2} \right]$$

$$+ \frac{m}{T} \left[- \frac{2}{\left(\dfrac{2\pi n}{T}\right)^2} \right]$$

$$A_n = \frac{m}{T} \left[\frac{2\, e^{-j2\pi n} - 2}{\left(\dfrac{2\pi n}{T}\right)^2} \right]$$

$$= \frac{2m}{T} \left[\frac{e^{-j2\pi n} - 1}{\left(\dfrac{2\pi n}{T}\right)^2} \right]$$

$$= \frac{2m}{T} \left[\frac{(-1)\,(e^{-j\pi n})\,(-e^{-j\pi n} + e^{j\pi n})}{\left(\dfrac{2\pi n}{T}\right)^2} \right]$$

$$= - \frac{2m}{T} [e^{-j\pi n}] \left[\frac{e^{j\pi n} - e^{-j\pi n}}{\left(\dfrac{2\pi n}{T}\right)^2} \right]$$

$$A_n = - \frac{4m}{T} [e^{-j\pi n}][j] \left[\frac{e^{j\pi n} - e^{-j\pi n}}{2j\left(\dfrac{2\pi n}{T}\right)^2} \right]$$

$$= - 2j \frac{m}{T} [e^{-j\pi n}] \frac{\sin \pi n}{\left(\dfrac{2\pi n}{T}\right)^2}$$

$$= -\frac{4m}{T} [je^{-j\pi n}] \frac{\sin \pi n}{\left(\frac{2\pi n}{T}\right)^2}$$

As can be seen, A_n will be zero for all integer values of n. Now we find the B_n coefficients.

$$B_n = \frac{2}{T} \int_0^T (mt) \sin \frac{2\pi nt}{T} \, dt$$

$$= \frac{2m}{T} \int_0^T t \sin \frac{2\pi nt}{T} \, dt$$

$$= \frac{2m}{T} \int_0^T t \left[\frac{e^{j\frac{2\pi nt}{T}} + e^{-j\frac{2\pi nt}{T}}}{2j} \right] dt$$

$$= \frac{2m}{Tj} \int_0^T t \, e^{j\frac{2\pi nt}{T}} \, dt$$

$$- \frac{m}{Tj} \int_0^T t \, e^{-j\frac{2\pi nt}{T}} \, dt$$

Again we apply the exponential integral,

$$\int xe^{ax} \, dx = \frac{e^{ax}}{a^2} (ax - 1)$$

Let $T = x$, $a = j\frac{2\pi n}{T}$

$$\frac{m}{Tj} \int_0^T t e^{j\frac{2\pi nt}{T}} \, dt$$

$$= \frac{m}{Tj} \left[\frac{e^{j2\pi n}}{-\left(\frac{2\pi n}{T}\right)^2} (j2\pi n - 1) \right]$$

$$- \quad \frac{1}{\left(\frac{2\pi n}{T}\right)^2} \quad (1) \Bigg]$$

Let $T = x, a - -j \dfrac{2\pi n}{T}$

Then

$$\frac{m}{-jT} \left[\frac{e^{-j2\pi n}}{-\left(\frac{2\pi n}{T}\right)^2} \; (-j2\pi n - 1) \right.$$

$$- \quad \frac{1}{\left(\frac{2\pi n}{T}\right)^2} \; (1) \Bigg]$$

$$= \frac{m}{jT} \left[\frac{e^{-j2\pi n}}{-\left(\frac{2\pi n}{T}\right)^2} \; (j2\pi n + 1) \right.$$

$$+ \quad \frac{1}{\left(\frac{2\pi n}{T}\right)^2} \; (1) \Bigg]$$

$$B_n = \frac{m}{jT} \left[\frac{j2\pi n \; (e^{j2\pi n} + e^{-j2\pi n})}{-\left(\frac{2\pi n}{T}\right)^2} \right.$$

$$+ \quad \frac{e^{j2\pi n}}{\left(\frac{2\pi n}{T}\right)^2} \quad - \quad \frac{1}{\left(\frac{2\pi n}{T}\right)^2}$$

$$- \quad \frac{e^{-j2\pi n}}{\left(\frac{2\pi n}{T}\right)^2} \quad + \quad \frac{1}{\left(\frac{2\pi n}{T}\right)^2} \Bigg]$$

$$B_n = \frac{m}{jT} \; \frac{(j2\pi n) \; (e^{j2\pi n} + e^{-j2\pi n})}{-\left(\frac{2\pi n}{T}\right)^2}$$

$$+ \; \frac{m}{jT} \left[\frac{e^{j2\pi n} - e^{-j2\pi n}}{\left(\dfrac{2\pi n}{T}\right)^2} \right]$$

$$B_n = \frac{m}{T} \left[\frac{e^{j2\pi n} + e^{-j2\pi n}}{-\dfrac{2\pi n}{T^2}} \right]$$

$$+ \; \frac{2jm}{jT} \left[\frac{e^{j2\pi n} - e^{-j2\pi n}}{2j\left(\dfrac{2\pi n}{T}\right)^2} \right]$$

$$B_n = m \, T \left[\frac{e^{j2\pi n} + e^{-j2\pi n}}{-2\pi n} \right]$$

$$\underbrace{+ \; \frac{2 \, m}{T} \; \frac{\sin 2\pi n}{\left(\dfrac{2\pi n}{T}\right)^2}}$$

This goes to zero for all n (integer values)

$$B_n = \frac{m \, T}{-\pi n} \left[\frac{e^{j2\pi n} + e^{-j2\pi n}}{2} \right]$$

$$= - \; \frac{m \, T}{\pi n} \cos 2\pi n$$

From the drawing, the slope was:

$$m = \frac{V}{T}$$

$$B_n = \frac{V}{-\pi n} \cos 2\pi n$$

For V = 0.4 volts we find the following:

$$\mathbf{B_n} \; \frac{\mathbf{n}}{1} \qquad \mathbf{B_n} \; \frac{0.4}{\pi} \; \cos 2\pi \; (1)$$

$$= -\frac{0.4}{\pi}$$

$$2 \qquad -\frac{0.4}{2\pi}$$

$$3 \qquad -\frac{0.4}{3\pi}$$

$$4 \qquad -\frac{0.4}{4\pi}$$

The Fourier Series is then given by:

$$A_O + \Sigma B_n \sin \frac{2\pi n T}{T}$$

$$= 0.2 - \frac{0.4}{\pi}\left[\sin 2\pi(10)t + \frac{1}{2} \sin 2\pi(20)t \right.$$

$$\left. + \frac{1}{3} \sin 2\pi(30)t + \frac{1}{4} \sin 2\pi(40)t \right]$$

Chapter 3

Problem 1.

First we must find the temperature:

$$
\begin{aligned}
T &= 273° + C \\
&= 273° + 17° = 290\,°K \\
E_n^2 &= 4k\,T\,R\,B \\
&= 4(1.38 \times 10^{-23})\,(2.9 \times 10^2)(3 \times 10^2)\,(6 \times 10^6) \\
&= 288 \times 10^{-13} = 28.8 \times 10^{-12} \text{ volts}^2 \\
E_n &= 5.37 \times 10^{-6} = 5.37\mu V
\end{aligned}
$$

Problem 2.

$$
\begin{aligned}
T &= 273° + 51° \\
&= 324\,°K
\end{aligned}
$$

$$E_n^2 = 4\text{k }T\,R\,B$$
$$= 4(1.38 \times 10^{-23})\,(3.24 \times 10^2)\,(3 \times 10^2)\,(6 \times 10^6)$$
$$= 322 \times 10^{-13} = 32.2 \times 10^{-12} \text{ volts}^2$$
$$E_n = 5.67\mu\text{V}$$

Problem 3.

$$R = \frac{E_n^2}{4k\;T\;B}$$

$$= \frac{(3 \times 10^{-7})^2}{4(1.38 \times 10^{-23})\,(2.9 \times 10^2)\,(7 \times 10^3)}$$

$$R = \frac{9 \times 10^{-14}}{112 \times 10^{-18}}$$

$$= \frac{9 \times 10^4}{112} = 803.6\ \Omega$$

Problem 4.

$$P_a = k\;T\;B$$
$$= (1.38 \times 10^{-23})\,(3 \times 10^2)\,(10^4)$$
$$= 4.14 \times 10^{-17} \text{ watts}$$

Problem 5.

$$\frac{E_{nt}}{Series} = \sqrt{En_1^2 + En_2^2}$$

$$= \sqrt{(50\ \mu\text{V})^2 + (70\ \mu\text{V})^2}$$

$$= \sqrt{7400}$$

$$= 86\ \mu\text{V}$$

For the parallel connection we derive the following equation:

$$E_{nt} = \sqrt{4k\;T\;B\;R_t}$$

$$R_t = \frac{R_1 R_2}{R_1 + R_2}$$

$$R_1 = \frac{E_{n1}^2}{4k\ T\ B}, \quad R_2 = \frac{E_{n2}^2}{4k\ T\ B}$$

$$R_t = \frac{\left(\dfrac{E_{n1}^2}{4k\ T\ B}\right)\left(\dfrac{E_{n2}^2}{4k\ T\ B}\right)}{\dfrac{E_{n1}^2}{4K\ T\ B} + \dfrac{E_{n2}^2}{4k\ T\ B}}.$$

$$E_{nt} = \sqrt{\frac{4k\ T\ B\left(\dfrac{E_{n1}^2}{4k\ T\ B}\right)\left(\dfrac{E_{n2}^2}{4k\ T\ B}\right)}{\dfrac{E_{n1}^2}{4k\ T\ B} + \dfrac{E_{n2}^2}{4k\ T\ B}}}$$

$$E_{nt} = \sqrt{\frac{(E_{n1}^2)\ (E_{n2}^2)}{E_{n1}^2 + E_{n2}^2}}\ \mu V$$

$$= \sqrt{\frac{(50)^2\ (70)^2}{(50)^2 + (70)^2}}\ \mu V$$

$$= \sqrt{\frac{(2500)\ (4900)}{7400}}\ \mu V$$

$$= \sqrt{1655.4}\ \mu V$$

$$= 40.7\ \mu V$$

Problem 6.

$$E_n^2 = \frac{k\ T}{C}$$

$$= \frac{(1.38 \times 10^{-23})\ (3 \times 10^2)}{4 \times 10^{-12}}$$

$$= 1.035 \times 10^{-23} \times 10^{14}$$

$$= 1.035 \times 10^{-9}$$

$$E_n = \sqrt{10.35 \times 10^{-10}} = 32.2 \ \mu V$$

Problem 7.

Noise equivalent circuits:

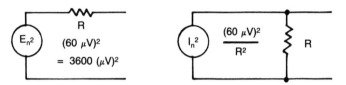

Problem 8.

$$R_t = \frac{R_1 \ (R_2 + R_3)}{R_1 + R_2 + R_3}$$

$$= \frac{10 \ K \ (20 \ K + 30 \ K)}{10 \ K + 20 \ K + 30 \ K}$$

$$= 8.33 \ k\Omega$$

$$E_n^2 = 4 \ k \ T \ R_t \ B$$

$$= 4 \ (1.38 \times 10^{-23}) \ (3 \times 10^2) \ (8.33 \times 10^3) \ (10^6)$$

$$= 137.9 \times 10^{-12}$$

$$E_n = 11.74 \ \mu V$$

Problem 9.

First we must find the temperature in degrees Kelvin. Using equation 3-6

$$T = \frac{5}{9} \ (F - 32) + 273$$

$$= \frac{5}{9} \ (50 - 32) + 273$$

$$= 283° \ K$$

Then we find f_x

$$
\begin{aligned}
f_x &= 0.15 \, k \, T \times 10^{34} \\
&= (0.15)(1.38 \times 10^{-23})(2.83 \times 10^2)(10^{34}) \\
&= 0.58 \times 10^{13} \\
&= 5.8 \times 10^{12} \text{ Hz}
\end{aligned}
$$

Problem 10.

For this problem we can derive a special formula:

$$
\begin{aligned}
E_n &= \sqrt{4 \, k \, T \, R \, B} \\
&= \sqrt{B} \, \sqrt{4 \, k \, T \, R} \\
&= \sqrt{B} \, \sqrt{4 \times 1.38 \times 10^{-23} \times 3 \times 10^2 \times 10^4} \\
&= \sqrt{B} \, \sqrt{16.56 \times 10^{-17}} \\
&= \sqrt{B} \, \sqrt{1.656 \times 10^{-16}} \\
&= \sqrt{B} \, (1.29 \times 10^{-8})
\end{aligned}
$$

We can make a table as shown below:

B Hz	E_n
100	$(\sqrt{100})(1.29 \times 10^{-8}) = 1.29 \times 10^{-7} = 0.129 \ \mu V$
200	.182
300	.223 μV
400	.258 μV
500	.288 μV

The curve has a nonlinear shape since E_n is proportional to $\sqrt{B}$ and not B.

Problem 11.

As temperature increases in a resistor, the thermal noise generated also increases. The noise voltage increases proportional to the temperature raised to the half power. As temperature lowers, so does the noise voltage.

Problem 12.

To prove that the area under the curve is unity, perform the following steps.

The total area under the probability curve write:

$$\int_{-\infty}^{\infty} P(x)dx = \int_{-\infty}^{\infty} \frac{1}{2 E_n^2} \epsilon^{\frac{-e_n^2}{2 E_n^2}} de_n$$

We now prepare the integral:

$$\frac{1}{} \int_{-\infty}^{\infty} \epsilon^{-\frac{e_n^2}{2 E_n^2}} de_n = \frac{1}{\sqrt{2 \pi E_n^2}}$$

$$\int_{-\infty}^{\infty} \epsilon^{-\left(\frac{e_n}{\sqrt{2 E_n}}\right)^2} de_n$$

This is because the integral is of the form

$$\int_{-\infty}^{\infty} \epsilon^{-x^2} \, dx$$

If we now multiply the differential den by $\dfrac{1}{\sqrt{2} E_n}$ and multiply the constant in front of the integral by $\sqrt{2} E_n$

$$\text{Area} = \frac{\sqrt{2 E_n^2}}{\sqrt{2 E_n^2}} \int_{-\infty}^{\infty} \epsilon^{-\left(\frac{e_n}{2 E_n}\right)^2} \frac{d_{en}}{\sqrt{2 E_n^2}}$$

Simplification gives:

$$\text{Area} = \frac{1}{\sqrt{\pi}} \int_{-\infty}^{\infty} \Sigma \; e^{-x^2} \, dx$$

The definite integral

$$\int_{-\infty}^{\infty} e^{-x^2} \, dx$$

Therefore,

$$\text{Area} = \frac{1}{\sqrt{\pi}} [\sqrt{\pi}] = 1$$

Problem 13.

To evaluate the noise across the circuit we first convert the center "Y" section to a delta network as shown in Fig. 3-13. The respective values are as follows:

$$R_A = \frac{R_1 R_2 + R_1 R_3 + R_2 R_3}{R_2}$$

$$= \frac{4(4) + (4)(2) + (4)(2)}{4}$$

$$= 8 \; \Omega$$

$$R_B = \frac{R_1 R_2 + R_1 R_3 + R_2 R_3}{R_1}$$

$$= \frac{32}{4}$$

$$= 8 \; \Omega$$

$$R_C = \frac{R_1 R_2 + R_1 R_3 + R_2 R_3}{R_3}$$

$$= \frac{32}{2}$$

$$= 16\ \Omega$$

Now the new network can be drawn as in Fig. 3-14, by placing this delta network back into the original network. We may now reduce the circuit to Fig. 3-15 because of the two delta networks in parallel and then obtain the circuit of Fig. 3-16. Another simplification yields that of Fig. 3-17. One final simplification yields the circuit of Fig. 3-19.

Computing the total thermal noise we have:

$$
\begin{aligned}
E_n^{\,2} &= 4\ k\ T\ R\ B \\
&= 4(1.38 \times 10^{-23})\ (3 \times 10^2)\ (3.89)\ (10^6) \\
&= 6.4.4 \times 10^{-14} \\
E_n &= 2.54 \times 10^{-7} \\
&= 0.2.54\ \mu V
\end{aligned}
$$

Chapter 4

Problem 1.

$$I_{ns} = \sqrt{2e\ I_{dc}\ B}$$

$$= \sqrt{2 \times 1.6 \times 10^{-19} \times 0.1 \times 3 \times 10^4}$$

$$= 0.96 \times 10^{-15} = 9.6 \times 10^{-16}\ \text{amps}^2$$

Problem 2.

$$E_{ns}^{\,2} = (I_{ns})^2\ (R)^2 = (9.6 \times 10^{-16})\ (4 \times 10^3)^2$$

$$= 153.6 \times 10^{-10}\ v^2$$

$$E_{ns} = 12.4 \times 10^{-5}$$

$$= 124\ \mu V$$

$$E_{nt}^{\,2} = 4k\ T\ R\ B$$

$$= (4)\ (1.38 \times 10^{-23})\ (2.9 \times 10^2)\ (4 \times 10^3)\ (3 \times 10^4)$$

$$= 192 \times 10^{-14} = 1.92 \times 10^{-12} \ v^2$$

$$E_{nt} = 1.386 \times 10^{-6} = 1.386 \ \mu V$$

$$E_n \ \text{total} = \sqrt{E_{ns}^2 + E_{nt}^2}$$

$$= \sqrt{153.6 \times 10^{-10} + .0192 \times 10^{-10}}$$

$$= 124 \ \mu V$$

Problem 3.

$$G_{I \ SHOT} = 2e \ I_{dc} = 2(1.6 \times 10^{-19}) \ (.1) = 3.2 \times 10^{-20} \ \frac{\text{amps}^2}{\text{Hz}}$$

$$G_{V \ SHOT} = 2e \ I_{dc} \ R^2 = (3.2 \times 10^{-20}) \ (4 \times 10^3)^2$$
$$= 51.2 \times 10^{-14} \ \text{volts}^2/\text{Hz}$$

$$G_{V \ THERMAL} = 4k \ T \ R = 4 \ (1.38 \times 10^{-23}) \ (2.9 \times 10^2) \ (4 \times 10^3)$$
$$= 64 \times 10^{-18} \ \text{volts}^2/\text{Hz}$$

$$G_{I \ THERMAL} = 4k \ T \ G = 4 \ (1.38 \times 10^{-23}) \ (2.9 \times 10^2) \frac{1}{(4 \times 10^3)}$$

$$= 4 \times 10^{-24} \ \text{amps}^2/\text{Hz}$$

The spectral density for the entire noise process will be about the same as that for the shot noise process.

Problem 4.

$$E_{no}^2 = (G) \ (Av)^2 \ (B)$$

$$= (5 \times 10^{-10}) \ (40)^2 \ (10^3)$$

$$= (5 \times 10^{-10}) \ (1.6 \times 10^3) \ (10^3)$$

$$= 8 \times 10^{-4} \ \text{volts}^2$$

$$E_{no} = 2.83 \times 10^{-2} = 28.3 \ \text{mV}$$

Problem 5.

$$R_{eq} = \frac{2.5}{gm}$$

$$= \frac{2.5}{4 \times 10^{-3}} = 625 \ \Omega$$

Problem 6.

$$R_{eq} = R_i + R_{eq} + \frac{R_L}{A^2}$$

$$= 10 \text{ k} + 2 \text{ k} + \frac{40 \text{ k}}{(5)^2}$$

$$= 12 \text{ k} + 1.6 \text{ k} = 13.6 \text{ k} \ \Omega$$

Problem 7.

$$R_{eqT} = R_{eq1} + \frac{R_{eq2}}{A_{v1}^2} + \frac{R_{eq3}}{A_{v1}^2 \bullet A_{v2}^2} + \frac{R_{eq4}}{A_{v1}^2 \bullet A_{v2}^2 \bullet A_{v3}^2}$$

$$= 50 \text{ k} \ \Omega + \frac{100 \text{ k} \ \Omega}{(5)^2} + \frac{200 \text{ k} \ \Omega}{(5)^2 (10)^2} + \frac{30 \text{ k} \ \Omega}{(5)^2 (10)^2 (20)^2}$$

$$= 50 \text{ k} \ \Omega + 4 \text{ k} \ \Omega + 0.8 \text{ k} \ \Omega + \frac{30 \text{ k} \ \Omega}{(25) (100) (400)}$$

$$= 54.8 \text{ k} \ \Omega + \frac{30 \text{ k} \ \Omega}{10^6}$$

$$= 54.8 \text{ k} + .03 \ \Omega$$

$$\cong 54.8 \text{ k}$$

Problem 8.

The signal-to-noise ratio for the circuit is as follows. The noise voltage due to the source is $E_n = \sqrt{4k\,T\,R_s\,B}$. This produces a voltage across R_g of

$$E_n \ \frac{R_g}{R_s + R_g}.$$

The mean square voltage is then

$$\frac{E_n^2 \, R_g^2}{(R_s + R_g)^2}.$$

The signal voltage across R_g is then

$$E_s \left(\frac{R_g}{R_s + R_g} \right)$$

The squared value of this would be

$$\frac{E_s^2 \, R_g^2}{(R_s + R_g)^2}$$

The power due to the noise is

$$\frac{E_s^2 \, R_g^2}{(R_s + R_g)^2} \left(\frac{1}{R_g} \right)$$

and the power due to the signal is

$$\frac{E_n^2 \, R_g^2}{(R_s + R_g)^2} \left(\frac{1}{R_g} \right)$$

The signal-to-noise ratio then becomes:

$$\frac{S}{N} = \frac{\dfrac{E_s^2 \, R_g^2}{(R_s + R_g)^2} \left(\dfrac{1}{R_g} \right)}{\dfrac{E_n^2 \, R_g^2}{(R_s + R_g)^2} \left(\dfrac{1}{R_g} \right)}$$

$$= \frac{E_s^2}{E_n^2} = \frac{E_s^2}{4k\,T\,R_s\,B}$$

Problem 9.

$$T = \frac{5}{9} (F - 32) + 273$$

$$= \frac{5}{9} (69 - 32) + 273$$

$$= \frac{5}{9} (37) + 273$$

$$= 293.6 \, °K$$

$$V_{no} = (\sqrt{4k \ T \ R_{eq} \ B}) \, (A_v)$$

$$= (\sqrt{4 \times 1.38 \times 10^{-23} \times 2.936 \times 10^2 \times 4 \times 10^3 \times 10^4}) \, (50)$$

$$= (\sqrt{64.83 \times 10^{-14}}) \, (50)$$

$$= (8.05 \times 10^{-7}) \, (50)$$

$$= 402.5 \times 10^{-7} = 40.25 \, \mu V$$

Problem 10.

Because R_{eq} does not really exist and cannot draw current.

Problem 11.

See Text.

Problem 12.

$$Ens \ = \sqrt{2k \ T \ B \ rd}$$

$$= \sqrt{2(1.38 \times 10^{-23}) \, (2.93 \times 10^2) \, (4 \times 10^3) \, (25)}$$

$$= \sqrt{808.68 \times 10^{-18}}$$

$$= \sqrt{8.0868 \times 10^{-16}} = 2.844 \times 10^{-8} = 28.44 \, nV$$

Problem 13.

$$\frac{S}{N}\bigg| dB = 10 \log \frac{S}{N} = 10 \log 12 = 10 \,(1.079) = 10.79 \text{ dB}$$

Problem 14.

See Text.

Chapter 5

Problem 1.

a) For the circuit composed of a capacitor and resistor in series:

$$Z = \frac{1}{jwc} + R$$

By inspection the real part is R.

b) For the circuit composed of a series LR branch in parallel with a capacitor we have,

$$Z = \frac{(R + jwL)\left(\dfrac{1}{jwc}\right)}{R + jwL + \dfrac{1}{jwc}}$$

$$= \frac{R + jwL}{jwcR - w^2 Lc + 1}$$

We can multiply both numerator and denominator by the denominator's conjugate:

$$Z = \frac{(R + jwL)}{(1 - w^2 Lc + jwcR)} \cdot \frac{(1 - w^2 Lc - jwcR)}{(1 - w^2 Lc - jwcR)}$$

$$Z = \frac{R - w^2 LcR - jwcR^2 + jwL - jw^3 L^2 c + w^2 L\, c\, R}{(1 - w^2 Lc)^2 + (wcR)^2}$$

We then gather the real terms in the numerator and divide by the denominator. The real part of the impedance is then:

$$\text{Real Part of } Z = \frac{R}{(1 - w^2 Lc)^2 + (wcR)^2}$$

c) For the circuit composed of a capacitor and resistor in parallel we have:

$$Z = \frac{R\,(\frac{1}{jwc})}{R + \frac{1}{jwc}} = \frac{R}{jwcR + 1}$$

$$Z = \frac{R}{(R + jwcR)}\,\frac{(1 - jwcR)}{(1 - jwcR)}$$

$$= \frac{R}{1 + (wcR)^2} - \frac{jwcR^2}{1 + (wcR)^2}$$

The real part of Z is:

$$\frac{R}{1 + (wcR)^2}$$

Problem 2.

We will not consider the noise from R. The noise bandwidth can be given by the formula:

$$B_{eq} = \frac{\pi}{2}\,\frac{1}{2\pi RC}$$

$$= \frac{1}{4\,R\,C}$$

$$= \frac{1}{4\,(10^4)\,(2 \times 10^{-10})}$$

$$= \frac{1}{8\,(10^{-6})} = 0.125\,(10^6) = 125\text{ kHz}$$

The noise output voltage can be given by:

$$E_{no} = \sqrt{G\,A_v^2\,B_{eq}}$$

$$= \sqrt{5 \times 10^{-8}\,(70)^2\,(1.25 \times 10^5)}$$

$$= \sqrt{(5)\,(4.9)\,(10^{-8})\,(10^3)\,(1.25)\,(10^5)}$$

$$= \sqrt{(30.625)\,(10^0)}$$

$$= 5.53 \text{ volts}$$

Problem 3.

From Chapter 2 we know:

$$\frac{E_o}{E_i} = \frac{1}{1 - w^2c^2R^2 + j3wcR}$$

$$\left|\frac{E_o}{E_i}\right| = \frac{1}{\sqrt{1 + 7w^2c^2R^2 + w^4c^4R^4}}$$

$$\left|\frac{E_o}{E_i}\right|^2 = \frac{1}{1 + 7w^2c^2R^2 + w^4c^4R^4}$$

We can ESTIMATE the integral to be of the form:

$$|A^{(f)}|^2 = \frac{1}{1 + w^4c^4R^4}$$

The noise equivalent bandwidth formula is given by:

$$B_{eq} = \frac{\displaystyle\int_0^\infty |A_{(f)}|^2\,df}{A_p^2}$$

Since $|A_{(o)}|^2 = A_p^2 = (1)^2$ we can then write:

$$B_{eq} = \frac{\displaystyle\int_0^\infty \frac{1}{1 + w^4 c^4 R^4}\, df}{(1)^2}$$

$$\int_0^\infty \frac{x^{m-1} dx}{1 + x^n} = \frac{\dfrac{\pi}{n}}{\sin \dfrac{m\,\pi}{n}}$$

By comparison, $w^4 c^4 R^4$ must equal x^n so n equals 4. Since there is no factor in the numerator we assume that it is raised to the zero power. Therefore $x^{m-1} = 1$ and $m = 1$.

Preparing the integral,

$$B_{eq} = \frac{1}{2\pi cR} \int_0^\infty \frac{df\,(2\pi cR)}{1 + (wcR)^4}$$

$$= \frac{1}{2\pi cR} \left[\frac{\dfrac{\pi}{4}}{\sin \dfrac{(1)\,(\pi)}{4}} \right]$$

$$B_{eq} = \frac{1}{2\pi cR} \left[\frac{0.785}{\sin 45°} \right]$$

$$= \frac{1}{2\pi cR} \left[\frac{0.785}{0.707} \right] = \left(\frac{1}{2\pi cR} \right) \quad (1.11)$$

Problem 4.

$$Z = R + \frac{\dfrac{1}{jwc}\left(R + \dfrac{1}{jwc} \right)}{\dfrac{1}{jwc} + R + \dfrac{1}{jwc}}$$

$$= R + \frac{R + \dfrac{1}{jwc}}{1 + jwcR + 1}$$

$$= R + \frac{jwcR + 1}{jwc - w^2c^2R + jwc}$$

$$= R + \frac{jwcR + 1}{j2wc - w^2c^2R}$$

To find the real part of the impedance we add R to the real part of the second term. We find the real part of the second term as follows:

$$\frac{jwcR + 1}{j2wc - w^2c^2R} = \frac{(jwcR + 1)}{(j2wc - w^2c^2R)} \frac{(-j2wc - w^2c^2R)}{(-j2wc - w^2c^2R)}$$

$$= \frac{2w^2c^2R - jw^3c^3R^2 - j2wc - w^2c^2R}{4w^2c^2 + w^4c^4R^2}$$

The real part of the term is:

$$\frac{w^2c^2R}{4w^2c^2 + w^4c^4R^2}$$

The real part becomes:

$$R_{e(Z)} = R + \frac{R}{4 + w^2c^2R^2}$$

Assuming a time constant cR = 1 we will have:

$$R_{e(Z)} = R + \frac{R}{4 + w^2}$$

w	Real Part
0	1.25R
1	1.2R
2	1.125R
3	1.077R
4	1.05R
10	1.0096R
∞	R

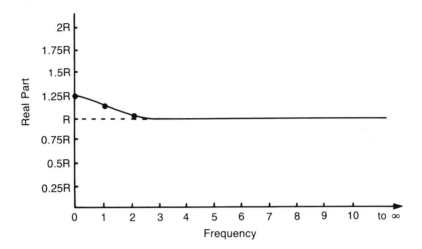

Problem 6.

A filter has a transfer function of:

$$H(w) = e^{-0.35\left(\frac{f}{B}\right)^2}$$

where B is the 3 dB bandwidth.

Determine the equation for the noise equivalent bandwidth.

$$B_{eq} = \frac{\displaystyle\int_0^\infty |A(f)|^2}{Ap^2}$$

$$= \frac{\displaystyle\int_0^\infty |e^{-0.35\left(\frac{f}{B}\right)^2}|^2 df}{Ap^2}$$

The maximum value or peak of the response occur at 0 hertz.

$$A(f) = e^{-0.35\left(\frac{O}{B}\right)^2} = 1$$

395

$$B_{eq} = \frac{\int_0^\infty e^{-0.7 \left(\frac{f}{B}\right)^2} df}{(1)^2}$$

Preparing the integral,

$$B_{eq} = \frac{B}{\sqrt{.7}} \frac{\int_0^\infty e^{-0.7 \left[\left(\frac{f}{B}\right)\right]^2} \frac{df\sqrt{.7}}{B}}{1}$$

$$= \frac{B}{\sqrt{0.7}} \int_0^\infty e^{-x^2} dx$$

$$\cong 1.2B \int_0^\infty e^{-x^2} dx$$

The integral portion is a definite integral that has a limit of

$$\frac{\sqrt{\pi}}{2}$$

Therefore, the noise equivalent bandwidth becomes:

$$B_{eq} = 1.2\,B \left[\frac{\sqrt{\pi}}{2}\right]$$

$$= 1.063\,B$$

Problem 7.

To evaluate this circuit's noise we go through the following procedure. The real part of the impedance is found by:

1. $Z = \dfrac{R(jw^2L)}{R + jwL}$

2. $Z = \dfrac{R(jwL)\,(R - jwL)}{(R + jwL)\,(R - jwL)}$

$$= \frac{jwLR^2}{R^2 + w^2L^2} + \frac{Rw^2L^2}{R^2 + w^2L^2}$$

3. The real part is:

$$\frac{Rw^2L^2}{R^2 + w^2L^2}$$

4. $E_n^2 = 4k\ T \displaystyle\int_0^\infty R_{e(f)}\ df$

$$= 4k\ T \int_0^\infty \frac{Rw^2L^2df}{R^2 + w^2L^2}$$

By examination of the integral we see that as f gets very large the integral goes to ∞. This simply means that as frequency rises the circuit response begins to get larger and the real part goes to R. The result is that theoretically the noise voltage is not limited by a noise equivalent bandwidth.

An alternate solution to this problem is as follows for a high pass RC circuit.

1. Find the transfer function:

$$\frac{E_o}{E_i} = \frac{i\ R}{i\left(\dfrac{1}{jwc} + R\right)}$$

$$= \frac{R}{\dfrac{1}{jwc} + R}$$

$$= \frac{jwc\ R}{1 + jwcR}$$

2. Find the magnitude of the transfer function:

$$\left|\frac{E_o}{E_i}\right| = \frac{(wcR)}{\sqrt{1 + (wcR)^2}}$$

3. Find the peak of the magnitude. For a high pass filter, this occurs at $w =$

$$\lim_{w \to \infty} \left|\frac{E_o}{E_i}\right| R = \frac{1}{\sqrt{\dfrac{1}{(wcR)^2} + 1}}$$

$$= \frac{1}{\sqrt{\frac{1}{\infty} + 1}}$$

$$= 1$$

4.

$$B_{eq} = \int_0^\infty \frac{|A(f)|^2}{Ap^2} \, df$$

$$B_{eq} = \frac{\int_0^\infty \dfrac{(wcR)^2 \, df}{1 + (wcR)^2}}{(1)^2}$$

$$= \frac{1}{2\pi CR} \int_0^\infty \frac{(wcR)^2 \, df(2\pi CR)}{1 + (wcR)^2}$$

As the frequency becomes very large, this integral moves towards infinity. This can be seen by

$$B_{eq} \cong \frac{1}{2\pi CR} \int_0^\infty dwCR$$

$$\text{since} \quad \frac{(wCR)^2}{1 + (wCR)^2} \quad \text{approaches unity.}$$

We therefore conclude that the noise equivalent bandwidth for a high-pass filter is infinity.

This is true for all types of high-pass filters. However, we must remember that all filters have some stray capacitance across the output terminals. This results in a roll-off at very high frequencies. This is another way of saying that at very high frequencies, an effective low-pass filter tends to cause the high-pass filter response to roll off. Therefore, a high-pass filter in the real world has a very large noise equivalent bandwidth instead of an infinite noise equivalent bandwidth.

Problem 8.

The total noise voltage due to the thermal effect is the same

for each network since the capacitors are equal.
The noise is given by:

$$E_n = \sqrt{\frac{k\,T}{C}}$$

Thus, there is no difference for either circuit for total thermal noise due to all frequencies.

The noise for a frequency range between 0 and 10 kHz could be calculated as follows:

$$E_n^{\ 2} = 4k\,T \int_0^\infty R_{e(f)}\,df$$

The real part of the impedance is given by

$$R_{e(f)} = \frac{R}{1 + (wCR)^2}$$

$$= 4\,kT \int_0^\infty \frac{R\,df}{1 + (wc\,R)^2}$$

$$= \frac{4\,kT}{2\,\pi c} \int_0^\infty \frac{df\,2\,\pi RC}{1 + (wc\,R)^2}$$

$$= \frac{4\,kT}{2\,\pi c} \int_0^\infty \frac{dwcR}{1 + (wc\,R)^2}$$

$$= \frac{4\,k\,T}{2\,\pi c} \left[\frac{1}{1} \arctan \frac{wc\,R}{1} \right]_0^B$$

$$= \frac{4\,kT}{2\,\pi c} \left[\arctan wc\,R \right]_0^B$$

$$= \frac{2\,kT}{\pi c} \left[\arctan 2\,\pi BCR \right]$$

For circuit #1

$$E_n^2 = \frac{2\,k\,T}{\pi\,(1 \times 10^{-9})}\,[\text{arctan } 6.28 \times 10^4 \times 10^{-9}\,(10^4)]$$

$$= \frac{2 \times 1.38 \times 10^{-23} \times 3 \times 10^2}{\pi\,(10^{-9})}\,[\text{arctan } 0.628]$$

$$= 2.64 \times 10^{-10}\,[\,32.1\,]$$

$$= 84.8 \times 10^{-10}\text{ volts}^2$$

$$E_n = 9.21 \times 10^{-5}\text{ volts or } 92.1\ \mu\text{V}$$

For circuit #2

$$E_n^2 = \frac{2\,k\,T}{\pi\,(1 \times 10^{-9})}\,[\text{arctan } 6.28 \times 10^4 \times 10^{-9}\,(10^5)]$$

$$= 2.64 \times 10^{-10}\,[\,80.95\,]$$

$$= 213.7 \times 10^{-10}\text{ volts}^2$$

$$E_n = 14.6 \times 10^{-5}\text{ volts or } 146\ \mu\text{V}$$

Problem 9.

The noise equivalent bandwidth for each of the circuits are as follows:

$$B_{el} = \frac{\pi}{2}\,f_c$$

$$= \frac{\pi}{2}\left(\frac{1}{2\pi RC}\right)$$

$$= \frac{1}{4\,RC}$$

$$= \frac{1}{4(10^4)\,(10^{-9})}$$

$$= \frac{10^5}{4}$$

$$= \quad 25 \text{ kHz}$$

$$B_{e2} \quad = \quad \frac{1}{4 \, RC}$$

$$= \quad \frac{1}{4(10^5) \, (10^{-9})}$$

$$= \quad \frac{10^4}{4}$$

$$= \quad 2.5 \text{ kHz}$$

Problem 10.

First we must determine the transfer function

$$H(jw) \quad = \quad \cfrac{\cfrac{SL\left(\cfrac{1}{CS}\right)}{SL + \cfrac{1}{CS}}}{R + \cfrac{SL\left(\cfrac{1}{CS}\right)}{SL + \cfrac{1}{CS}}}$$

$$= \quad \cfrac{\cfrac{\cfrac{L}{C}}{SL + \cfrac{1}{CS}}}{R + \cfrac{\cfrac{L}{C}}{SL + \cfrac{1}{CS}}}$$

$$= \quad \cfrac{\cfrac{L}{C}}{R\left(SL + \cfrac{1}{CS}\right) + \cfrac{L}{C}}$$

$$= \quad \cfrac{SL}{R(S^2LC + 1) + SL}$$

$$\frac{SL}{S^2LCR + R + SL}$$

$$= \frac{SL}{S^2LCR + SL + R}$$

We now set up the integral. First we must find the value of the function when the peak occurs. Substitute jw for w and we have:

$$\frac{jwL}{-w^2LCR + jwL + R}$$

$$A_{(f)} = \frac{jwL}{(R - w^2LCR)^2 + R^2}$$

By inspection, the denominator will minimize when

$$w = \frac{1}{LC.}$$

At such a frequency the function becomes:

$$|A_{(f)}| = \frac{\frac{1}{\sqrt{LC}} \cdot L}{R}$$

$$= \frac{1}{R}\sqrt{\frac{L}{C}}$$

This then makes the function have a maximum squared peak value of

$$\left(\frac{1}{R}\sqrt{\frac{L}{C}}\right)^2$$

The noise equivalent bandwidth will then be:

$$B_{eq} = \frac{\displaystyle\int_0^\infty \frac{(wL)^2 \, df}{(R - w^2LCR)^2 + R^2}}{\left(\frac{1}{R^2}\frac{L}{C}\right)}$$

$$= \frac{\displaystyle\int_0^\infty \frac{(wL)^2 \, df}{2R^2 - 2w^2LCR^2 + w^4L^2C^2R^2}}{\dfrac{1}{R^2} \dfrac{L}{C}}$$

$$= \frac{\displaystyle\int_0^\infty \frac{(wL)^2 \, df}{2 - 2w^2LC + w^4L^2C^2}}{\dfrac{L}{C}}$$

$$= \frac{\displaystyle\int_0^\infty \frac{C}{L} \; \frac{(wL)^2 \, df \times 2\pi \sqrt{LC}}{2 - 2w^2LC + w^4L^2C^2}}{\dfrac{L}{C}}$$

We can now estimate the problem by

$$B_{eq} = \frac{1}{2\pi\sqrt{LC}} \frac{\sqrt{2}}{2} \int_0^\infty \frac{\left(\dfrac{x}{\sqrt[4]{2}}\right)^2 \, dx}{1 + \left(\dfrac{x}{\sqrt[4]{2}}\right)^4}$$

$$= \frac{1}{2\pi\sqrt{LC}} \frac{1}{\sqrt{2}} \left(\frac{\pi/4}{\sin 3/4 \ \pi}\right)$$

$$= \frac{1}{2\pi\sqrt{LC}} \frac{\pi}{4}$$

Problem 11.

$$Z = \frac{\left(R1 + \dfrac{1}{jwC} + jwL\right)R2}{R1 + R2 + \dfrac{1}{jwC} + jwL} \qquad \text{but } R1 = R2 = R_{-j}$$

$$= \frac{\left(R + \dfrac{1}{jwC} + jwL\right)R}{2R + \dfrac{1}{jwC} + jwL}$$

$$= \frac{(jwCR + 1 - w^2LC)R}{j2wCR + 1 - w^2LC}$$

$$= \frac{jwCR^2 + R - w^2LCR}{j2wCR + 1 - w^2LC}$$

Conjugate

$$\frac{(jwCR^2 + R - w^2LCR)}{(j2wCR + 1 - w^2LC} \qquad \frac{(-j2wCR + 1 - w^2LC}{(-j2wCR + 1 - w^2LC)}$$

$$= \frac{2w^2C^2R^3 + jwCR^2 - jw^3C^2R^2L - j2wCR^2 + R - w^2LCR + j2w^3LC^2R^2 - w^2LCR + w^4L^2C^2R}{4w^2C^2R^2 + 1 - 2w^2LC + w^4L^2C^2}$$

The real part is then extracted:

$$R_{e(Z)} = \frac{2w^2C^2R^3 + R - 2w^2LCR + w^4L^2C^2R}{1 + 4w^2C^2R^2 - 2W^2LC + w^4L^2C^2}$$

$$= \frac{R\,[1 + w^2C^2R^2 - 2w^2LC + w^4L^2C^2]}{1 + w^2[4C^2R^2 - 2LC] + w^4L^2C^2}$$

Problem 12.

$$H_{(f)} = 10e^{-0.35\,f/B}$$

404

1. $|A_{(f)}| = |H_{(f)}| = 10e^{-0.35\,f/B}$

2. $|A_{(f)}|^2 = 100e^{-0.7\,f/B}$

3. $A_p = 10$

4. $A_p^2 = 100$

5.

$$B_{eq} = \frac{\displaystyle\int_0^\infty |A_{(f)}|^2\,df}{A_p^2}$$

$$= \frac{\displaystyle\int_0^\infty 100e^{-0.7\,f/B}\,df}{100}$$

$$= \int_0^\infty e^{-0.7\,f/B}\,df$$

$$B_{eq} = \left[\frac{e^{-0.7\,f/B}}{\dfrac{-0.7}{B}} \right]_0^\infty$$

$$= \frac{B}{-0.7}\,e^{-\infty} - e^0$$

$$= [-1.43B]\,[0 - 1]$$

$$= 1.43\,B$$

This simply states that the noise equivalent bandwidth is 1.43 multiplied by the 3 dB bandwidth of the filter.

Problem 13.

1. Find the area under the curve. The area can be broken into two triangles and one rectangle. The respective areas are:

$$A_1 = 1/2\ b\ h$$
$$= 1/2(4000\ \text{Hz})\ (600)$$
$$= 1/2(4 \times 10^3)\ (6 \times 10^2)$$
$$= 1.2 \times 10^6\ \text{Hz}$$
$$A_2 = 1/2\ b\ h$$
$$= 1/2(2000\ \text{Hz})\ (600)$$
$$= 6 \times 10^5\ \text{Hz}$$
$$A_3 = b\ h$$
$$= (6 \times 10^3)\ (6 \times 10^2)$$
$$= 3.6 \times 10^6\ \text{Hz}$$

2. Find the total area.

$$A_T = A_1 + A_2 + A_3$$
$$= 1.2 \times 10^6 + 0.6 \times 10^6 + 3.6 \times 10^6$$
$$= 5.4 \times 10^6$$

3. Find B_{eq}

$$B_{eq} = \frac{\text{Total Area}}{A_p^{\ 2}}$$
$$= \frac{5.4 \times 10^6}{6 \times 10^2}$$
$$= 9000\ \text{Hz}$$

Problem 14.

Find the noise equivalent bandwidth of the circuit that has the transfer function of:

$$H_{(f)} = 10e^{-f/B}$$

1. $A_{(f)} = H_{(f)} = 10e^{-f/B}$

2. $|A_{(f)}| = |H_{(f)}| = 10e^{-f/B}$

3. $|A_{(f)}|^2 = 100e^{-2\ f/B}$

4. $B_{eq} = \dfrac{\displaystyle\int_{o}^{\infty} |A_{(f)}|^2\ df}{A_p^{\ 2}}$

5. $A_p = 10$

Index

Edited by Roland S. Phelps

6.　　$A_p^2 = 100$

7.　　$B_{eq} = \dfrac{\displaystyle\int_0^\infty 100e^{-2\,f/B}\,df}{100}$

8.　　$B_{eq} = \displaystyle\int_0^\infty e^{-2f/B}\,df$

$$= \dfrac{e^{-2f/B}}{\dfrac{-2}{B}}\,\Bigg|_0^\infty$$

$$= -\dfrac{B}{2}\,[e^{-\infty} - e^0]$$

$$= -\dfrac{B}{2}\,[0 - 1]$$

$$= \dfrac{B}{2}$$

Chapter 6

Problem 1.

$$F = \frac{N_o}{N_i G} = \frac{70\ \mu w}{(15\ \mu w)(3)} = 1.56$$

Problem 2.

$$F = 1 + \frac{N_n}{N_i G} = 1 + \frac{40\ \mu w}{(2\ \mu w)(4)} = 6$$

Problem 3.

$$N_{oMIN} = N_i G + (2\ \mu w)(4) = 8\ \mu w$$

Problem 4.

$$F_{dB} = 10 \log 1.2 = 0.79\ \text{dB}$$

$$F_{dB} = 10 \log 10 = 10 \text{ dB}$$
$$F_{dB} = 10 \log 20 = 13 \text{ dB}$$

Problem 5.

$$N_o = N_n + N_iG = 15 \ \mu w + 4 \ \mu w(3) = 27 \ \mu w$$

$$F = \frac{N_o}{N_iG} = \frac{27 \ \mu w}{12 \ \mu w} = 2.25$$

Problem 6.

$$F_1 = 6, \ G_1 = 10, \ F_2 = 10, \ G_2 = 3$$

$$F = F_1 + \frac{F_2 - 1}{G_1} = 6 + \frac{10 - 1}{10} = 6.9$$

Problem 7.

$$F = F_1 + \frac{F_2 - 1}{G_1} + \frac{F_3 - 1}{G_1G_2} + \frac{F_4 - 1}{G_1G_2G_3}$$

$$= 10 + \frac{14}{3} + \frac{4}{3 \bullet 6} + \frac{19}{3 \bullet 6 \bullet 10}$$

$$= 10 + 4.66 + 0.222 + 0.106$$

$$= 15$$

Problem 8.

$$G = 12, \ G = G_1G_2 \ \therefore \ G_1 = G_2 = \sqrt{12}$$

$$F = 3 = F_1 + \frac{F_2 - 1}{G_1} \qquad \text{But, } F_1 = F_2$$

$$3 = F_1 + \frac{F_1 - 1}{\sqrt{12}}.$$

$$3\sqrt{12} = F_1 \sqrt{12} + F_1 - 1$$
$$10.2 = 3.4 \, F_1 + F_1 - 1$$
$$11.2 = 4.4 \, F_1$$
$$F_1 = 2.55 = F_2$$

Problem 9.

$$G_1 = 7, \, F_1 = 10, \, G_2 = 6, \, F_2 = 10$$

$$F = \frac{G_1 F_1 + G_2 F_2}{G_1 + G_2} = \frac{7(10) + 6(10)}{13} = 10$$

Problem 10.

$$F = 1 + \frac{T_{eq}}{T_o} = 1 + \frac{29\,^\circ\text{K}}{290\,^\circ\text{K}} = 1.1$$

Problem 11.

$$T_{eq} = T_{eq1} + \frac{T_{eq2}}{G_1} + \frac{T_{eq3}}{G_1 G_2}$$

$$= 678\,^\circ\text{K} + \frac{325\,^\circ\text{K}}{10} + \frac{470\,^\circ\text{K}}{(10)(20)}$$

$$= 678\,^\circ\text{K} + 32.5\,^\circ\text{K} + 2.35\,^\circ\text{K}$$

$$= 712.85\,^\circ\text{K}$$

Problem 12.

Use $G_1 = G_2 = \sqrt{17}$,

$$T_{eq} = 150\,^\circ\text{K} + \frac{300\,^\circ\text{K}}{4.2} = 150\,^\circ\text{K} + 71.4\,^\circ\text{K}$$

$$= 221.4\,^\circ\text{K}$$

$$F = 1 + \frac{T_{eq}}{T_o} = 1 + \frac{221.4}{290}$$

$$= 1 + 0.762$$

$$= 1.762$$

Problem 13.

$$F = 1 + \frac{T_{eq}}{T_o} = \frac{G_1 F_1 + G_2 F_2}{G_1 + G_2}$$

$$F_1 = 1 + \frac{T_{eq1}}{T_o} = 1 + \frac{150}{290} = 1.517$$

$$F = 1 + \frac{T_{eq2}}{T_o} = 1 + \frac{300}{290} = 2.03$$

$$F = \frac{8.5\,(1.517) + 8.5\,(2.03)}{17}$$

$$= \frac{12.9 + 17.3}{8.4} = 3.58$$

$$T_{eq} = T_o\,(F-1) = 290\,(3.58 - 1)$$

$$= 290\,(2.58)$$

$$= 748° \text{ K.}$$

Chapter 7

Problem 2.

From Example 2-8 in Chapter 2, the rms value of a square wave is the same as its peak value. Therefore, a square wave with a peak value of 20 volts has an rms value of 20 volts.

A full-wave average responding voltmeter would respond to the full-wave average of the waveform, which in this case would also be 20 volts. Since the meter would be calibrated for a sine-

wave, which has a form factor of 1.11, the meter would read (E_{av}) (1.11) = 22.2 volts. The meter should read 20 volts. Therefore the meter reading should be multiplied by $\dfrac{20}{22.2}$ or 0.9. The correction factor would then be 0.9.

Problem 3.

If the square wave of Problem 2 was applied to a true rms meter, the meter would read 20 volts.

Problem 4.

The rms noise voltage would be $\sqrt{0.78 \times 10^{-6}}$ = 0.883 × 10^{-3} = 0.883 mV.

a) If this voltage was applied to a full-wave average responding voltmeter the meter would read

$$\frac{0.883 \text{ mV}}{1.15} = 0.768 \text{ mV}.$$

b) If this voltage was applied to a true rms voltmeter, the meter would read 0.883 mV.

Chapter 8

Problem 2.

Spot noise figure refers to the noise figure at a particular point in the frequency spectrum for a network. Average noise figure would relate to the entire frequency spectrum. For example, below we have some theoretical information concerning noise figure.

Frequency Range	Noise Figure
10 - 110 Hz	10
110 - 200 Hz	15
200 - 300 Hz	20

The spot noise figure for a particular frequency, say 250 Hz,

would be 20. While the average noise figure for the entire frequency spectrum could be approximated as $10 + 15 + 20$ all divided by 3.

Problem 4.

$$E_{no1} = 20 \ \mu V$$

$$E_{no2} = 3 \ E_{no1} = 3 \ (20 \ \mu V) = 60 \ \mu V$$

$$R_{eq} = (R_a) \frac{(20 \ \mu V)^2}{(60 \ \mu V)^2 - (20 \ \mu V)^2}$$

$$= 5 \ k \left(\frac{400}{3600 - 400} \right)$$

$$= 5 \ k \left(\frac{400}{3200} \right)$$

$$= 5 \ k \left(\frac{1}{8} \right)$$

$$= 625 \ \Omega$$

Problem 5.

Use Equation 8-6

$$B_{eq} = \frac{\text{Summation of individual areas under square of voltage gain}}{(A_p)^2}$$

$$= \frac{50,000 \ Hz}{(20)^2}$$

$$= \frac{50,000}{400}$$

$$= 125 \ Hz$$

Problem 6.

$$E_{neq} = \frac{V_{no}}{A_v} = \frac{50 \ \mu V}{70} = 0.714$$

Problem 7.

$$F = 20 \, I_{bc} \, R_g \quad \text{Note: } R_g \text{ is made equal to } R_{in}$$

$$= 20 \, (4 \times 10^{-3}) \, (300)$$

$$= 24$$

Problem 8.

$$T_h = 800\,°K, \; T_c = 100\,°K$$

$$M = \left(\frac{E_{no2}}{E_{no1}}\right)^2 = \left(\frac{80}{40}\right)^2 = 4$$

$$T_{eq} = \frac{T_h - M \, T_c}{M - 1} = \frac{800 - 4(100)}{4 - 1}$$

$$= \frac{400}{3} = 133.3\,°K$$

$$= 1 + \frac{T_e}{T_o} = 1 + \frac{133.3}{290} = 1.46$$

Chapter 9

Problem 1.

$$\frac{S_o}{N_o} = \frac{S_i}{N_i} \left[\frac{(2m^2)}{2 + m^2} \right]$$

$$\frac{S_i}{N_i} = \frac{S_o}{N_o} \left[\frac{2 + m^2}{2m^2} \right]$$

$$= 200 \left[\frac{2 + (.3)^2}{2(.3)^2} \right]$$

$$= 200 \left[\frac{2 + 0.09}{0.18} \right]$$

$$= 200 \left[\frac{2.09}{0.18} \right]$$

$$= 200 \left[11.6 \right]$$

$$= 2322$$

Problem 2.

Since noise figure is given by the formula:

$$F = \frac{\dfrac{S_i}{N_i}}{\dfrac{S_o}{N_o}} \, ,$$

we could use the following equation:

$$\frac{\dfrac{S_i}{N_i}}{\dfrac{S_o}{N_o}} = \frac{1}{m^2} + \frac{1}{2}$$

Since the upper limit on the noise figure is infinity this would occur when m = 0. The minimum noise figure would occur when m = 1. The noise figure would then become:

$$F = \frac{1}{(1)^2} + \frac{1}{2} = 1.5$$

Problem 3.

$$\frac{S_o}{N_o} = 2 \left(\frac{S_i}{N_i} \right) = 2\,(67) = 134$$

414

Problem 5.

$$\frac{\dfrac{S_o}{N_o}\bigg|FM}{\dfrac{S_o}{N_o}\bigg|AM} = 3\,(m_f)^2 \ = \ 3\left(\frac{\Delta f}{fm}\right)^2$$

$$= \ 3\left(\frac{25 \text{ kHz}}{15 \text{ kHz}}\right)^2$$

$$= \ 3\,(2.78)$$

$$= \ 8.33$$

Problem 6.

If amplitude of modulation doubled, then Δf would double, becoming 50. This would cause m_f to become 3.33 and the improvement to become 33.33.

Problem 8.

$$v_q = \frac{v_q}{\sqrt{12}}$$

$$= \frac{2}{\sqrt{12}}$$

$$= \ 0.577 \text{ volts.}$$

Chapter 10

Problem 1.

$$V_{no}^{\;2} = 4\,k\,T\,B\left(\frac{R_s R_g}{R_s + R_g} + R_{eg}\right)A_v^{\;2} + 4\,k\,T\,B\,R_L$$

$$= 4 \; k \; T \left(\left[\frac{R_s R_g}{R_s + R_g} + R_{eg} \right] A_v^{\;2} + R_L \right) B$$

$$4 \; k \; T = 4 \times 1.38 \times 10^{-23} \times 3 \times 10^2$$

$$= 16.56 \times 10^{-21} = 1.656 \times 10^{-20}$$

USE NOISE EQUIVALENT BANDWIDTH

$$B_e = \frac{\pi}{2} \quad f_c = \frac{1}{4 \; RC}$$

$$R = \frac{r_d \; R_L}{r_d + R_L} = 5 \; \text{k}\Omega$$

$$B_e = \frac{1}{4 \; (5 \times 10^3) \; (1 \times 10^{-10})}$$

$$= \frac{1}{20 \times 10^{-7}} = \frac{1}{2 \times 10^{-6}}$$

$$= 500 \; \text{kHz}$$

$$\left(\frac{R_s \; R_g}{R_s + R_g} + R_{eq} \right) A_v^{\;2} + R_L =$$

$$\left(\frac{(2\text{k}) \; (1\text{m})}{2\text{k} + 1\text{m}} + 4 \; \text{k} \right) \left(\frac{-40 \; (10\text{k})}{10\text{k} + 10\text{k}} \right)^2 + 10\text{k} \cong$$

$$(2\text{k} + 4\text{k}) \; (400) \; + 10\text{k} \cong$$

$$2{,}410{,}000 \; \Omega$$

$$V_{no}^{\;2} = 1.656 \times 10^{-20} \; (2.41 \times 10^6) \; (5 \times 10^5)$$

$$= 19.95 \times 10^{-9}$$

$$= 1.995 \times 10^{-8} \; v^2$$

$$V_{no} = 1.412 \times 10^{-4} v$$

Since $R_g >> R_s$

$$F = 1 + \frac{R_{eq}}{R_s} + \frac{R_L}{R_s A_v^2}$$

$$= 1 + \frac{4k}{2k} + \frac{10k}{2k(-20)^2}$$

$$= 1 + 2 + .0125$$

$$= 3.0125$$

Problem 2.

Since $R_g >> R_s$

$$F = 1 + \frac{R_{eq}}{R_s} + \frac{R}{R_s}$$

$$= 1 + \frac{4k}{2k} + \frac{10k}{2k}$$

$$= 1 + 2 + 5$$

$$= 8$$

Problem 3.

$$F = 1 + \frac{R_s}{R} + \frac{R_{eq}}{R_s}\left(\frac{R + R_s}{R}\right)^2$$

$$= 1 + \frac{2k}{10k} + \frac{4k}{2k}\left(\frac{10k + 2k}{10k}\right)^2$$

$$= 1 + 0.2 + 2\,(1.44)$$

$$= 4.08$$

Problem 4.

$$R_{so} = \sqrt{(R_{eq}) \, (R_g)}$$

$$= \sqrt{(4k) \, (1000k)}$$

$$= 63.24 \ k\Omega$$

$$\left(\frac{N_s^2}{N_p}\right) = \frac{R_{so}}{R_s} = \frac{63.24k}{2k}$$

$$= 31.6$$

$$\frac{N_s}{N_p} = \sqrt{31.6}$$

$$= 5.62 \ \text{or} \ \frac{562}{100}$$

Problem 5.

$$F_M = 1 + \frac{v_b}{R_s} + \frac{1}{\sqrt{h_{FE}}}$$

$$= 1 + \frac{200}{100} + \frac{1}{\sqrt{70}}$$

$$= 1 + 2 + .119$$

$$\cong 3.12$$

Problem 8.

The noise gain is:

$$A_n = 1 + \frac{Z_f}{R_1}$$

with no compensation

$$Z_f = R_2$$

$$A_n = 1 + \frac{100k}{4k}$$

$$= 26$$

The compensation resistor is:

$$\frac{R_1 R_2}{R_1 + R_2} = \frac{(4k)(100k)}{4k + 100k}$$

$$= 3.84 \text{ k}\Omega$$

Appendix B

Comprehensive Noise Test

The following represents a comprehensive exam in electrical noise. At the completion of studying this book, you should be able to answer the majority of the following questions. Allow yourself three hours to take the exam. When you are finished, you can check your answers at the end of the book.

1. Electrical noise can be broken into three main categories. What are their names?

2. For each of the following indicate whether it is an example of erratic, man-made or circuit noise.

    a. electrical noise from thunderstorms.
    b. electrical noise from an electric razor.
    c. electrical noise from a car tire.
    d. electrical noise from a car ignition system.
    e. popcorn noise from a transistor.
    f. thermal noise from a resistor.
    g. electrical interference from the sun and stars.
    i. shot noise in a transistor.
    j. electrical noise from a computer or calculator.
    k. electrical noise from a dc motor.

3. Sketch a model of a voltage induced in a wire due to light-

ning. The maximum voltage is 1000 volts.

4. Make a spectral density plot of both pink and white noise.

5. Write the statistical term that best represents the following noise terms:

a. square of noise voltage rms level
b. rms value of noise

6. Suppose that the spectral density of a pink noise process is 2×10^{-5} volts^2/ Hz at 1 Hz. What is the mean square output voltage in a frequency range of interest from 700 to 3500 Hz.

7. Find the rms and average level for the following waveform.

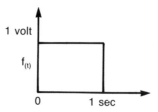

8. Determine the Fourier transform for the following waveform.

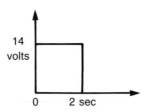

9. Convolve the following functions using graphical methods.

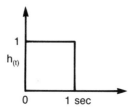

10. Draw the autocorrelation function for the following wave-form.

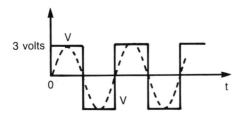

11. Find the amplitude of the eleventh harmonic in the following waveform.

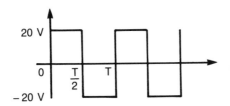

12. Determine the shot and thermal noise associated with the following circuit:

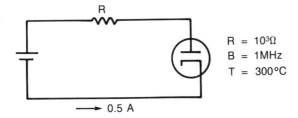

$R = 10^3\Omega$
$B = 1MHz$
$T = 300°C$

0.5 A

13. If the shunt capacity of a resistor is 9 pF, what is the total rms noise voltage?

14. The highest frequency where a certain thermal noise source can be considered flat is 5×10^{11} Hz. Compute the temperature of the noise source.

15. A shot noise spectral density of 6×10^{-14} volts^2 per hertz is applied to the input of an amplifier with a voltage gain of 1000 and a bandwidth of 400 Hz. The output noise voltage is 100 mV.

Determine the bandwidth.

16. The triode has a noise equivalent resistance of 670 ohms. What is the transconductance.

17. Find the total noise equivalent resistance of the following network: $R_{eq} = 10$ k, $R_{eq}^2 = 120$ k, $R_{eq}^3 = 1$ MΩ, $A_{v1} = 4$, $A_{v2} = 10$.

18. Find the signal to noise ratio across this resistor.

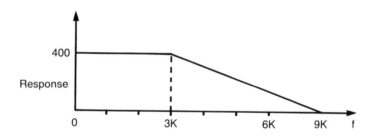

$$e_s = 10 \text{ volts} \qquad e_n = .01 \text{ volts}$$

19. Determine the noise equivalent bandwidth for the following network.

20. Given the following data, compute the overall noise figure as well as the overall noise equivalent temperature.

$$\boxed{F_1 = 10} \longrightarrow \boxed{F_2 = 12} \longrightarrow T_o = 290°K$$

21. Explain in words what is meant by a noise figure of six.

22. A noise process having a value of 7×10^{-9} volts squared is applied to an average responding voltmeter (full wave). What is the voltage that the meter should read?

23. Determine the noise equivalent bandwidth using the following information.

1) Area under the voltage gain squared characteristic: 8.6×10^6 Hz.

2) Voltage gain peak equals $\sqrt{4300}$.

24. A noise figure measurement is made with a vacuum tube noise generator. The input resistance of the system is 300 ohms. The noise figure is 30. What is the plate current reading when the output noise power is doubled?

25. An FM system has a frequency deviation of 30 kHz and modulation frequency of 5 kHz. What is the output signal-to-noise ratio improvement over a comparable AM system with a modulation index of one?

26. Determine the noise figure for the circuit shown below.

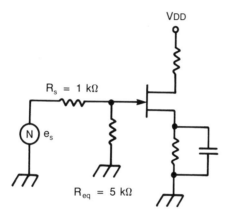

Appendix C

Answers to the Comprehensive Electrical Noise Test

1. erratic, man-made, spontaneous fluctuation.

2. a) erratic b) man-made c) erratic d) man-made e) circuit noise f) circuit noise g) erratic i) circuit noise j) circuit noise k) man-made

3. In this waveform, the voltage rises to 1000 volts in about 1 - 2 μsec. It then decays in an exponential manner to 500 volts in an additional 48.8 μsec. The drawing is as follows:

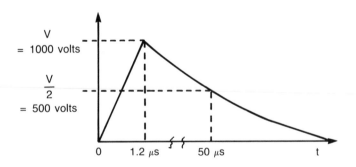

4.

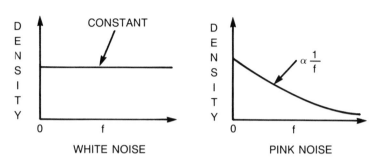

WHITE NOISE   PINK NOISE

5. The square of noise voltage rms level could be described as a variance. The rms level of noise could be described as a standard deviation.

6.

$$E_n^2 = \int_{f_1}^{f_2} k \left(\frac{1}{f}\right) df$$

$$= k \int_{f_1}^{f_2} f^{-1} df$$

$$= k \left[ln\ f \right]_{f_1}^{f_2}$$

$$E_n^2 = 2 \times 10^{-5} \left[ln\ f \right]_{700}^{3500}$$

$$= 2 \times 10^{-5} \left[ln\ \frac{f_2}{f_1} \right]$$

$$= 2 \times 10^{-5} \left[ln \right]_{700}^{3500}$$

$$= 2 \times 10^{-5} \, [ln \, 5]$$

$$= 2 \times 10^{-5} \, [1.609]$$

$$= 3.219 \times 10^{-5} \, \text{volts}^2$$

7.

$$E_{AV} = \frac{\text{Area Under Waveform}}{\text{Baseband}}$$

$$= \frac{(5 \text{ volts}) (2 \text{ msec.})}{8 \text{ msec.}}$$

$$= 1.25 \text{ volts}$$

$$E_{rms} = \sqrt{\frac{\text{Area Under Function Squared}}{\text{Period}}}$$

$$= \sqrt{\frac{(5 \times 5) (2 \text{ msec})}{8 \text{ msec}}}$$

$$= \sqrt{\frac{25}{4}}$$

$$= \sqrt{6.25 \text{ volts}^2}$$

$$= 2.5 \text{ volts.}$$

8.

$$F_{(w)} = \int_{-\infty}^{\infty} f(t) \, e^{-jwt} dt$$

$$= \int_{\frac{-r}{2}}^{\frac{r}{2}} V \, e^{-jwt} dt$$

$$= Vr \, \frac{\sin \frac{wr}{2}}{\frac{wr}{2}}$$

$$F_{(w)} = 14 \, (2) \, \frac{\sin w \left(\frac{2}{2}\right)}{w \left(\frac{2}{2}\right)}$$

$$= 28 \frac{\sin w}{w}$$

9. To convolve the functions, fold $h_{(t)}$ back on $f_{(t)}$ as shown:

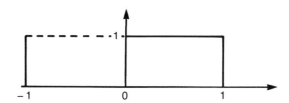

Slide $-f_{(t)}$ into $h_{(t)}$ and compute area overlap
The result should be as follows:

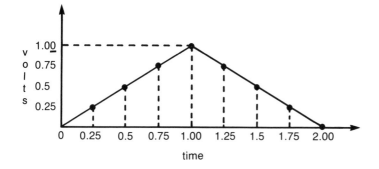

10.

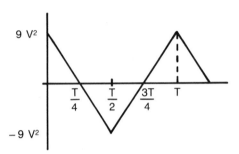

11. The eleventh harmonic would have an amplitude of

$$\frac{4 \text{ V}}{11 \ \pi} = \frac{4(20)}{11 \ (\pi)} = 2.316 \text{ volts.}$$

12.

$$E_{n \ shot} = \sqrt{2eI_{dc}B \ R^2}$$

$$= \sqrt{2(1.6 \times 10^{-19} \ (0.5) \ (10^6) \ (10^3)^2}$$

$$= \sqrt{3.2 \times 10^{-19} \ (0.5) \ (10^{12})}$$

$$= \sqrt{1.6 \times 10^{-7}}$$

$$= \sqrt{0.16 \times 10^{-6}}$$

$$= 0.4 \ mV.$$

$$E_{n \ Thermal} = \sqrt{4k \ T R \ B}$$

$$= \sqrt{4 \times 1.38 \times 10^{-23} \ (3 \times 10^2) \ (10^3) \ (10^6)}$$

$$= \sqrt{16.56 \times 10^{-12}}$$

$$= 4.07 \ \mu V.$$

13.

$$E_n = \sqrt{\frac{k\,T}{c}}$$

$$= \sqrt{\frac{(1.38 \times 10^{-23})(3 \times 10^2)}{9 \times 10^{-12}}}$$

$$= \sqrt{0.46 \times 10^{-9}}$$

$$= \sqrt{4.6 \times 10^{-10}}$$

$$= 2.145 \times 10^{-5}$$

$$= 21.45 \ \mu V.$$

14.

$$f_x = 0.15\,k\,T \times 10^{34}\text{Hz}$$

$$5 \times 10^{11} = 0.15\,(1.38 \times 10^{-23})(10^{34})T$$

$$5 \times 10^{11} = 2.07 \times 10^{10}T$$

$$T = 24.15°\,K$$

15.

$$E_n^2 = G\,(A)^2\,(B)$$

$$(0.1)^2 = (6 \times 10^{-14})(10^3)^2(B)$$

$$B = \frac{(.01)}{(6 \times 10^{-14})(10^6)}$$

$$= \frac{1 \times 10^{-2}}{6 \times 10^{-8}}$$

$$= 0.167 \times 10^6$$

$$= 167 \text{ kHz.}$$

16.

$$R_{eq} = \frac{2.5}{gm}$$

$$gm = \frac{2.5}{R_{eq}}$$

$$= \frac{2.5}{670}$$

$$= 3731 \ \mu V$$

17.

$$R_{eqT} = R_{eq1} + \frac{R_{eq2}}{A_{V1}^2} + \frac{R_{eq3}}{A_{V1}^2 \times A_{V2}^2}$$

$$= 10 \ k + \frac{120 \ k}{(4)^2} + \frac{1 \ M\Omega}{(4)^2 \ (10)^2}$$

$$= 10 \ k + 7.5 \ k + 0.625 \ k$$

$$= 18.125 \ k \ \Omega$$

18.

$$\left(\frac{e_s^2}{e_n}\right) = \left(\frac{10}{0.01}\right)^2 = \frac{100}{1 \times 10^{-4}}$$

$$= 1 \times 10^6$$

19.

$$B_{eq} = \frac{\text{Area under voltage gain squared}}{\text{peak voltage gain squared.}}$$

$$= \frac{400 \text{ V}^2 \times 3\,k + \frac{1}{2} \times 400 \times 6\,k}{400}$$

$$= 6 \text{ kHz.}$$

20.

$$F = F_1 + \frac{F_2 - 1}{G_1} + \frac{F_3 - 1}{G_1 G_2}$$

$$= 10 + \frac{12 - 1}{5} + \frac{4}{5(2)}$$

$$= 10 + \frac{11}{5} + \frac{4}{10}$$

$$= 10 + 2.2 + 0.4$$

$$= 12.6$$

$$T_{eq} = (F - 1)\, To$$

$$= (12.6 \bullet 1)(290°)$$

$$= 11.6\,(290°)$$

$$= 3364°\,k.$$

21. A noise figure of six means that the circuit in question is producing five times as much noise as the input noise multiplied by the voltage gain squared.

22.

$$V_{noise} = 1.15\, V_{meter}$$

$$V_{meter} = \frac{V_{noise}}{1.15}$$

$$= \frac{7 \times 10^{-9}}{1.15}$$

$$= 6.086 \times 10^{-9} \text{ volts}$$

23.

$$B_{eq} = \frac{\text{Area under (voltage gain)}^2}{\text{(voltage gain peak)}^2}$$

$$= \frac{8{,}600{,}000 \text{ Hz}}{4300}$$

$$= \frac{8.6 \times 10^6}{4.3 \times 10^3}$$

$$= 2 \times 10^3$$

24.

$$F = 20 \, I_{dc} R_g$$

$$30 = 20(I_{dc})\,(300)$$

$$1.5 = I_{dc}\,(300)$$

$$I_{dc} = 5 \; mA.$$

25.

$$\left.\frac{S_o}{N_o}\right|_{FM} = 3 \; (m_f)^2 \left.\frac{S_o}{N_o}\right|_{AM}$$

$$\left.\frac{S_o}{N_o}\right|_{FM} = 3 \left(\frac{\Delta f}{fm}\right)^2 \left.\frac{S_o}{N_o}\right|_{AM}$$

$$= 3 \left(\frac{30}{5}\right)^2 \left.\frac{S_o}{N_o}\right|_{AM}$$

$$= 3 \ (6)^2 \ \left. \frac{S_o}{N_o} \right| \text{AM}$$

$$= 108 \ \left. \frac{S_o}{N_o} \right| \text{AM}$$

26.

$$F \cong 1 + \frac{R_{eq}}{R_s} = 1 + \frac{5 \ k}{1 \ k} = 6$$